Overselling the Web?

Overselling the Web?

Development and the Internet

Charles Kenny

Viva Books

New Delhi | Mumbai | Chennai | Kolkata | Bengaluru | Hyderabad | Kochi | Guwahati

First Indian Edition 2008
Reprinted 2010

VIVA BOOKS PRIVATE LIMITED

- 4737/23 Ansari Road, Daryaganj, New Delhi 110 002
  E-mail: vivadelhi@vivagroupindia.net, Tel. 42242200
- 76, Service Industries, Shirvane, Sector 1, Nerul, Navi Mumbai 400 706
  E-mail: vivamumbai@vivagroupindia.net, Tel. 27721273, 27721274
- Jamals Fazal Chambers, 26 Greams Road, Chennai 600 006
  E-mail: vivachennai@vivagroupindia.net, Tel. 28290304, 28294241
- B-103, Jindal Towers, 21/1A/3 Darga Road, Kolkata 700 017
  E-mail: vivakolkata@vivagroupindia.net, Tel. 22836381, 22816713
- 7, Sovereign Park Apartments, 56-58, K. R. Road, Basavanagudi, Bangalore 560 004
  E-mail: vivabangalore@vivagroupindia.net, Tel. 26607409, 26607410
- 101-102 Mughal Marc Apartments, 3-4-637 to 641, Narayanguda, Hyderabad 500 029
  E-mail: vivahyderabad@vivagroupindia.net, Tel. 27564481, 27564482
- Beevi Towers, First Floor, SRM Road, Kaloor, Kochi 682 018, Kerala
  E-mail: vivakochi@vivagroupindia.net, Tel: 0484-2403055, 2403056
- 232, GNB Road, Beside UCO Bank, Silpukhuri, Guwahati 781 003
  E-mail: vivaguwahati@vivagroupindia.net, Tel: 0361-2666386

www.vivagroupindia.com

Published by arrangement with

Lynne Rienner Publishers, Inc.
1800 30th Street, Suite 314
Boulder, Colorado 80301
USA

ISBN : 978-81-309-0965-3

Published by Vinod Vasishtha for Viva Books Private Limited, 4737/23, Ansari Road, Daryaganj, New Delhi-110 002. Printed and bound by Anand Sons, Delhi - 110 092.

S0000 / S0300 / M0000 / M0203

Contents

Preface

The Nation is now in its 95th month of continuous economic advance. Both in strength and length, this prosperity is without parallel in our history. We have steered clear of the business-cycle recessions which for generations derailed us repeatedly from the path of growth and progress.

THE ABOVE QUOTE COMES from the Report of the US Council of Economic Advisers. It is, as it happens, not from a report to Bill Clinton toward the end of his term, but from one to Richard M. Nixon in 1969.[1] Many others were terribly optimistic regarding the effects of the 1946 Employment Act—the perhaps surprising object of the paen above.

In the 1970s, the term *New Economy* was coined to describe a dawning era of stable growth based on the emergence of a service economy, globalization, the rise of small companies, and corporate restructuring. In 1986, *Fortune* hailed "America's New Economy," and *Business Week* argued that the United States was "undergoing the most revolutionary economic change in a century."[2] Sadly, as we now know, the predictions of an end to the cyclical economy in the late 1990s were as inaccurate as those in the mid-1980s and the early 1970s.

Confidence in the latest version of the new economy has been shaken in the United States by the deflation of the Internet bubble. However, in much of the rest of the world—and perhaps particularly in the developing world—the language of the Internet revolution is alive and well. For many in development circles, we are still, by and large, in the Internet bubble phase, and government programs designed to foster that revolution are ramping up.

These programs must be good ones, because governments in developing countries have very limited resources: little cash, few skilled personnel, and modest stocks of equipment. At the same time, they face enormous and urgent development challenges. Over 1 billion people subsist on less than a dollar a day, and nearly half of the world's population on less than $2 a day. Twenty-seven percent of children under age five in the developing world are malnourished, and 20 million people have died of AIDS since 1986.[3] If the Internet has a role to play in meeting those challenges, it needs to be elaborated. At the same time, if there are limits to that role, they need to be clarified as well.

In this book I argue that a process of rationalization regarding the use of the Internet may need to take place in the development field, akin to what took place in the late 1990s in the United States. Just as in the United States, companies and individuals in developing countries will find that, many times, investment in the Internet makes sense and the technology really is likely to be transformative. But just as US investors have begun asking where the revenues will come from to repay their investment, the development community should begin asking: Are the benefits of this particular Internet investment worth the costs?

It is vital to move the analysis onto that level before billions of aid dollars and government budgets in less developed countries (LDCs) are spent wiring remote villages. In the United States, when investors irrationally bid up stock prices before the bubble burst, most of those affected were, by global standards, the very rich. If Tanzania, or Bangladesh, or India starts misallocating resources by investing in the wrong technologies in the wrong place, those who suffer will be the global poor. Avoiding an Internet bubble in development investments matters far more than failed attempts to control the stock bubble in the United States.

The Internet presents real opportunities to people in developing countries. In this book I discuss some of those opportunities and how they are being grasped. Indeed, one of the incredibly impressive things about Internet diffusion in the developing world is its rapidity—how fast entrepreneurs in Africa, Asia, and Latin America have seen the potential of the new technology and exploited it. Already by 2005, there were more Internet users in the developing than in the developed world. It may be that a silent minority or majority of development thinkers—passive skeptics—actually underestimate the impact of the Internet on economic development. The social impact of the Internet has already been large (there are 5 million members of one Indian online matchmaking service alone, for example) and doubtless will get larger.[4] That impact may well have dramatic economic consequences for developed and developing countries alike. Nonetheless, it is clear that the Internet cannot solve every problem, and there are significant downsides to the new technology from a developing country standpoint, which is the second focus of this book.

Various coauthors from previous collaborations supplied a number of the ideas and evidence that I appropriate in the book, including Richard Heeks, Christine Qiang, Juan Navas-Sabater, Carsten Fink, Emmanuel Forestier, Jeremy Grace, Rym Keremane, and Isabel Neto. (Of course, they may not necessarily agree with my arguments.) I would also like to thank the following journals as I reuse ideas and data that appeared as part of articles originally published by them: *International Journal of Educational Development, Oxford Development Studies, Telecommunications Policy, Development Policy Review, info,* and *Communications and Strategies.*[5] I would like to thank Beth Partin, Bjorn Wellenius, Michael Best, and the anonymous reviewers for very helpful and constructive comments and editorial suggestions on earlier drafts. Any remaining errors and all views are, of course, mine alone.

Finally, thanks to my wife, Pamela, who put up with many gripes and excuses based on editing and rewriting. The Internet in development was in some ways responsible for us meeting, and so, regardless of concerns raised in this book, I know it can be a force for very good things.

—Charles Kenny

Notes

1. Quoted in Galbraith 1989: 255.
2. Madrick 2001.
3. Figures from http://www.developmentgoals.org/.
4. Halewood and Kenny 2005.
5. Grace and Kenny 2003; Kenny 2003; Forestier, Grace, and Kenny 2001; Kenny 2002a; Kenny 2002b; and Kenny 2001.

1 — Will the Internet Change the World?

THE WORLD OF INFORMATION processing and communications has seen dramatic changes since the mid-1970s. Twenty-five years ago, no one had been saved by using their mobile phone to call for help after a car crash, and no one had been killed in a crash with someone who was talking on the phone rather than thinking about driving. Twenty-five years ago, young cyber-geeks typed out BASIC programs on machines with a whole 1k of memory, no mouse, and no graphical user interface (read: Windows).[1] Only a little before that, the height of interactive gaming was Pong.

Over the intervening years, the cost of voice transmission circuits has dropped by a factor of 10,000, and computing power per dollar invested has risen by a factor of 10,000. The effects of these changes have already begun to work their way into the economy. They have digitized and globalized the world of banking and investment, which sees over 1 trillion electronic financial transactions daily. E-commerce, broadly defined to include proprietary networks such as electronic data interchange (EDI), topped $1 trillion in the United States in 2000, according to the Census Bureau.[2]

These changes have, with some justification, produced material for countless editorials in the outmoded paper-based press, countless books in the outmoded paper-based publishing world, and countless commentaries on the outmoded analog TV, celebrating the new world in which we will live.

At the extreme, these writings have suggested millenarian change. George Gilder, an author and visionary, wrote in the *Wall Street Journal* on December 31, 1999, that the Internet would be the savior of religion:

> Now, at the turn of the new millennium, in a further unfolding of the overthrow of matter, we are moving into an industrial era based on photons, totally massless bearers of electromagnetic energy: light. . . . With any technology

> that will change the world so radically as the Internet . . . religious wars are important and inescapable. . . . The twentieth century has been an era when an atheistic belief in the ultimacy of matter and the triviality of man led to the horrors of Nazism, Communism, and an epoch of total war. Now sweeping through the global economy, the overthrow of matter will unleash an undertow of religious belief that will make the new millennium a time of awakening to the oceanic grandeur and goodness of the universe.[3]

But the idea that the Internet will change the world—developed and developing alike—in truly dramatic ways is not limited to the US religious right. A wide swath of businesspeople, senior officials, and development professionals agree. Commonly, they compare the Internet with the Industrial Revolution. Internet-driven changes "will transform our society over the next century as significantly as the two industrial revolutions"—one of railways and factories, the second of the internal combustion engine, electricity, synthetic chemicals and the automobile—argues Michael Dertouzos in *What Will Be*. The information marketplace, as he termed it, will establish itself "solidly and rightfully as the Third Revolution in modern human history. It is big, exciting and awesome." Similarly, in March 1999 Jack Welch, at that point the chief executive officer (CEO) of General Electric, also suggested that the arrival of the Internet was "the single most important event in the US economy since the Industrial Revolution," and Alan Greenspan, chairman of the Federal Reserve, was talking of a revolution that "has altered the structure of the way the American economy works."[4]

Thomas Friedman, author of *The Lexus and the Olive Tree* and op-ed writer for the *New York Times*, is another firm believer in the global Internet revolution. In the period between January 1999 and June 2001, he published pieces that mentioned the Internet and globalization thirty-four times. His arguments run as follows:

"If you don't think it's a new world, think again," he says. "We are now in a period of radical change, possibly more sweeping and complex than any period since 1776–1789." Technology, he argues, "is shrinking the world from a size medium to a size small." Because of the Internet, "you now have to think globally. You have to think about your customers as global, your competitors as global, your readers as global, your suppliers as global and your partners as global"—and that is happening "whether we like it or not."

At the same time, it "turns out that the real secret of success in the information age is what it always was: fundamentals—reading, writing and arithmetic, church, synagogue and mosque, the rule of law and good governance." Indeed, these basics have become even more important. "Just when the developing world is coming to really grasp that it has no choice but to get itself ready to climb aboard this train . . . the train is going to get faster—not slower—as the developing world moves toward Internet-based commerce, communication and learning systems. What's worse, no one can slow the train down, because the world economy today is just like that Internet: everybody is connected but nobody is in charge."

One reason that Friedman is such a believer in the Internet is that he argues it "super-empowers" people. He points to an upstart online book retailer challenging Amazon.com that got 142,000 hits after it was mentioned in one of his columns and argues that "with $100,000, I could start an Amazon.com tomorrow." At a slightly larger scale, America Online's (AOL's) global member services center in the Philippines, which employs 900 people, "is but one example of a simple truth—that the best hope for alleviating poverty around the world is through more globalization, not less." He notes that Jody Williams organized a global ban on landmines using e-mail. He approvingly quotes a nongovernmental organization (NGO) official who argues that "thanks to globalization and the Internet, power is now much more diffused, global companies are now much more exposed, and organizations like ours much better positioned to offer solutions."

But a number of countries aren't reacting well to the process of Internet-catalyzed globalization. "The failure of many nations to master modernity," Friedman warns, "is producing a lot of unemployed and angry young people in those countries, combined with the spread of new information technologies, which are super-empowering these angry people in ways that not only threaten the stability of the states they live in but also enable them, as individuals, to threaten America." Friedman cites the Love Bug virus created by two Filipinos that caused $10 billion in damage. He says that the Love Bug episode is as emblematic of the new world order as the Cuban missile crisis was of the Cold War.

The only way to avoid countries harboring "the rogue unemployed" and posing a threat to the world at large is "nation building—helping others restructure their economies and put in place decent, non-corrupt government," Friedman argues. Once again, however, the Internet appears to be making this process far easier. In the Middle East, he notes, "the Internet and globalization are acting like nutcrackers to open societies and empower Arab democrats with new tools. . . . In the 20th century, the Arab states thrived by developing oil. In the 21st century they will thrive only if they develop their people, and the only way to do that is by democratizing. The battle is on, and thanks to the Internet and globalization the Arab democrats finally have some artillery."[5]

So, according to Friedman, the Internet is a huge force for globalization and for development. Countries that get left behind will find themselves in ever-greater trouble. But the Internet is also a huge force for ensuring countries get what they need to stay abreast—reading, writing, and arithmetic; church, synagogue, and mosque; and the rule of law and good governance.

From the academic community, Nicholas Negroponte of the MIT Media Lab broadly agrees with Friedman's line of argument, suggesting that the Internet will promote rapid convergence of incomes:

> Developing nations will leapfrog the telecommunications infrastructures of the First World and become more wired (and wireless). We once moaned about the demographics of the world. But all of a sudden we must ask ourselves: Considering two countries with roughly the same population, Germany and Mexico, is it really so good that less than half of all Germans are under 40 and so bad that more than half of all Mexicans are under 20? Which of those nations will benefit first from "being digital"?[6]

Negroponte also believes that the Internet will foster world peace by breaking down national borders.[7] He suggests that, twenty years from now, children who are used to finding out about other countries through the click of a mouse "are not going to know what nationalism is."[8]

Belief in the transformative power of the Internet as a force for development is also widespread in industry. On October 16, 2000, a meeting of luminaries from the fields of development and information technology gathered in Seattle to discuss the global digital divide. The gathering was attended by figures such as Jeff Bezos (founder of Amazon), Eric Benhamou (chairman of 3Com), and Vint Cerf (founding father of the Internet).[9]

The group discussed a raft of private and public initiatives to achieve digital development—not least the proposed $1 billion Hewlett Packard World e-Inclusion program (since scaled back). "Providing the technologies to connect people everywhere is a vital social mission," said William Plummer, a vice president at Nokia. The sentiment echoed that of an interview given by John Chambers, the CEO of Cisco Systems, a couple of months before. "For the first time, we have the chance to address global poverty," he argued. "We have the chance to reeducate the majority of the world. If you are in Africa and your teachers are dying of AIDS, there is no way you can educate your population without this [Internet] capability. So it changes everything."[10]

Meanwhile, the heads of government of the Group of Eight (G8) dedicated their 2000 meeting in Okinawa to the Internet's role in development.[11] The G8's Charter on the Global Information Society declared: "Information and Communications Technology (IT) is one of the most potent forces in shaping the twenty-first century. . . . IT is fast becoming a vital engine of growth for the world economy. . . . Enormous opportunities are there to be seized and shared by us all."[12]

Members of the G8 went on to make dramatic promises of assistance to the cause of the Internet in development (although few were fully funded). Most spectacular was a $10 billion program announced by the host nation, Japan. At the same time the G8 created the Digital Opportunity Task Force to advise in the process of overcoming the "digital divide."

Among development officials, Mark Malloch Brown, then head of the United Nations Development Programme (UNDP), suggested in 2000 that "by eliminating space and time [the Internet] gives us an unprecedented means of overcoming two of the root causes of extreme poverty—ignorance and isolation.

. . . it will for the first time allow many poor and isolated groups to become part of the global community."[13] As a result of interest from leaders in business, government, and the donor community, capitals from London to Canberra began developing strategies to integrate information and communication technologies (ICTs) into their aid work, while capitals from Brasilia to Ulan Bator developed strategies in part to attract those dollars.

For developing countries, the first stage was frequently an "e-readiness assessment," in which the country ranked itself according to measures such as the number of phone lines per person, the number of government websites, and the number of information technology (IT) graduates produced each year.[14] Countries were also ranked by others, including the McConnell International survey *Risk E-Business: Seizing the Opportunity of Global E-Readiness,* the Mosaic Group's *Global Diffusion of the Internet Project,* the World Information Technology and Services Alliance *International Survey of E-Commerce,* the Center for International Development and Conflict Management's *Negotiating the Net,* the International Telecommunications Union's *Digital Access Index,* Orbicom's *Monitoring the Digital Divide and Beyond,* and Ohio State University's *Cyber-Space and Post-Industrial Transformations: A Cross-National Analysis of Internet Development.*[15] Bridges, a South African think tank, put together a helpful collection of such studies, which noted that fifty-five countries had been ranked by more than five different exercises.[16]

Hitching development to the Internet bandwagon, followed by the creation of numerous e-development assessment methodologies, gave rise to what might be termed the Okinawa consensus. It can be summarized as follows: The Internet and related technologies present a significant opportunity for developing countries to improve their growth prospects. Indeed, the Internet may be a "leapfrog" technology—one that creates an opportunity for developing countries to catch up economically with the industrial world. It is a powerful tool to improve government service delivery, education, and income-earning opportunities even for the world's poorest people. Given that, poor country governments (in partnership with the private sector and with the help of donors) need to dedicate significant resources to expanding the use of the Internet, especially in government and education and to reach the poor. In addition, Internet industries can be promoted by establishing technology parks, and Internet use can be increased by putting computers in libraries and building stand-alone Internet access points.

This book examines the Okinawa consensus, the policies suggested, the rationale behind them, and which ones might make sense. Suffice it here to say that we were and are acting in a relative vacuum of knowledge regarding those policies. The technology of the Internet is very young and its use in the poorest, or least-developed, countries even more recent. We can have, perforce, very little evidence on the effectiveness of Internet policies or of the Internet in development more generally.

As of December 2005, for example, there had not been survey-based, academically rigorous study of the economic impact of an Internet access program in any developing country. At the macroeconomic level, we have even less empirical knowledge about the impact of the Internet on economic growth than we do about older technologies or policies. For example, the Federal Reserve Bank of Cleveland published a cross-country study that used a novel approach to predict the impact of the Internet on global growth rates. It estimated the impact of cross-country Internet use in 1999 on gross domestic product (GDP) growth between 1974 and 1992, suggesting that 100 percent Internet usage in 1999 would have caused about 4 percent additional economic growth each year in the 1970s and 1980s.[17] Concerns over the direction of causality in this study surely throw some considerable doubt on the Reserve Bank's conclusions. Nonetheless, that such studies are published by the US Federal Reserve system is a sign of the difficulty economists face in this area.

Policy formulation based on anecdote is perhaps the inevitable result of the Internet's novelty. And even though anecdotal experiences are a perfectly valid form of evidence, no worse than the theoretically suspect statistical work based on weak data often presented as the alternative, such an approach to policy formulation needs to be used with caution.

By way of illustration, a recent study of research on the effectiveness of criminal justice measures found that research design had a large impact on the findings obtained.[18] Studies that lacked a well-chosen control group or a temporal sequence between cause and effect, studies that lacked random samples, and particularly studies that used anecdotes (rather than the results of experiments) as evidence all found larger positive and lower negative impacts of the intervention under review than did randomized studies with carefully designed control groups. The authors have little new to say as to why studies in which researchers have a larger degree of freedom as to what counts as evidence tend to be more positive, but one reason comes to mind—researchers with a vested interest (consciously or unconsciously) biasing results to fit their hopes for new programs.

With regard to use of the Internet in developing countries, the dominant source of evidence is preliminary "success stories" involving the use of the Internet for a particular application in a particular milieu, the story often as not told by the person who championed this use in the first place. Competitions such as the much-hyped Stockholm Challenge, which gives out prizes to managers of ICT projects from around the world who write in to describe what their ICT application has achieved, positively encourage this form of evidence creation. Such anecdotal evidence of cases in which the Internet is a cost-effective method of meeting an information need in LDCs is all that can be used to formulate global policy proposals at the macro level calling for widespread subsidies for access and to ICT firms. Yet there are numerous reasons that proponents might quite unintentionally bias their anecdotal reporting in favor of a positive outcome or generalize from a very particular set of circumstances that led to success.

To take an example of one real success story, Sakshi, an NGO advocating for women's rights in India, has long lobbied for legislation to protect women from sexual harassment. To bolster their efforts, they requested research support over the Internet from international women's groups. With the help of this support, Sakshi succeeded in having landmark sexual harassment legislation passed in India in 1997.[19]

The Internet *can* be a powerful tool to help improve social conditions, then. Yet it is a big step to argue, based on such stories, that the Internet will be a powerful force for growing social inclusion worldwide. To go from anecdote to universal application, we have to ask how replicable the anecdotal situation is—will it be repeated all over the world? In how many cases will the Internet make the difference between social exclusion and inclusion?

It is worth revisiting the Sakshi story in this regard. Knowledge of sexual harassment cases and legal precedents that was collected over the Internet was one element required in order to get the antiharassment laws approved. But a lot else was needed besides: Sakshi staff skilled in legal argument and in drafting writs, a receptive legislature with an institutional setup that allowed for NGO submissions, and an international network of support that provided good advice at low or no cost. At a more basic level, Sakshi needed money to pay staff who were literate, spoke English, and had computer skills; it needed electricity and telephone connections—the list goes on. But it is an important list because many people in the world are not literate, let alone computer-savvy, English-speaking legal experts. Worldwide, most people do not have access to the physical networks required for Internet use or the human networks providing high-quality free advice that is relevant to their concerns. Many people in the world live in countries where the supreme court would not be receptive to NGO interventions and unwilling to pass sexual harassment laws. The Internet was doubtless a useful tool in the fight to pass sexual harassment legislation in India, but it could only be effective in the role of knowledge conduit that led to change if a number of other factors came into play.

In addition to our limited information regarding replicability of success stories because of the context-specific nature of their success, a system that relies on self-reporting of stories and rewards positive results reporting over negative stories inevitably fosters selection bias. For example, a recent survey of garment and horticulture firms in three developing countries found that, of the sample of seventy-seven firms, all with access to the Web, only one trader in one company suggested that his use of the web had "revolutionised" the way he did business.[20] If anyone from the survey was to apply to the Stockholm Challenge or get their story in *Wired*, it would almost inevitably be that one person who had reported revolutionary results.

A focus on self-reported success stories still at the very early stages of development might well allow projects that turn out to be unsustainable to be put forward as models for replication. Take the 2001 Stockholm Challenge award-winning Gyandoot project in the Indian state of Madhya Pradesh, where more than 40 percent of the population live below the poverty line. Gyandoot was designed to bring forty-four e-government services to the people through thirty-nine telecenters connected via an intranet, including information on agricultural prices, a public complaint line regarding government services, and application facilities for various certificates.

The kiosks were set up at an average cost of approximately $1,500 each, and that expenditure has brought some positive returns. The Stockholm Challenge judges praised it as "a unique government-to-citizen Intranet project . . . with numerous benefits to the region, including a people-based self-reliant sustainable strategy 'Gyandoot' is recognized as a breakthrough in e-government."[21]

Nonetheless, a 2002 in-depth analysis found significant problems with implementation. A survey of the telecenters found 36 percent closed during the (working hours) survey visit. Of open centers, 35 percent had no electricity at the time of the visit, and 50 percent had no regular intranet connectivity. Average revenues per center per month averaged only $3—clearly far too low to suggest sustainability. Perhaps most damning, the survey team found an average of just over one user per center to interview, after searching for users both in the centers and in nearby community meeting areas. Survey results suggest an average of perhaps nineteen users per month per center (largely from upper income brackets) for the e-government services offered by the program.[22] It is too early to say that this experiment involving e-government for the poorest was a total failure, but 2001 was far too early to brand it an unqualified success.

Compare our patchy knowledge of the role of the technology of the Internet in development to the large and rich literature on the role of the technology of voice telephony. There again we have numerous anecdotal examples, but also survey-based as well as cross-country studies of the micro- and macroeconomic impact of telecommunications rollout. This literature clearly suggests impacts such as improved farm and nonfarm income earning opportunities, reduced prices for purchased goods, and reasonably compelling evidence of some type of link to economic growth.[23] We have had a good deal longer to evaluate the telephone (the technology has been around for a century), but this evaluation, based on surveys and statistics as well as stories, provides a far stronger basis for policy recommendations.

Again, there is a place for self-reported success stories as part of a spectrum of evidence regarding the development effectiveness of the Internet, but there are limits to how far such an approach can take us. The development community has begun to move beyond that approach—a number of careful studies and independent analyses are under way, and a few are already completed—but most of the existing evidence, I would argue, is still of the "success story" type. Given the problems of relying on anecdote alone, looking at any other sources of evidence that might bring some clarity to the potential role of the Internet in development is important.

First, it might be worth reviewing the theory and evidence regarding the role of other technologies, or technology in general, in promoting economic growth. It appears that there are no general rules on what causes growth when and where—as we have already seen regarding Sakshi, context matters. At the same time, it appears that the role played by technology in economic growth is very large pretty much everywhere. These two statements can both be true at the same time because, as I show in Chapter 2, economists have a very broad definition of technology. It is defined by economists as "total factor productivity," technology is anything that creates (or destroys) economic value that isn't capital or labor. It could include a new machine design, new law, or new accounting technique. Chapter 2 goes on to discuss what we know about technology and growth. In short, if technology is important to growth, rich countries must have more of it than poor countries, and given that the world still has a number of poor countries, moving technology around must be difficult. If the Internet resembles the average technology, then, it may have a role in growth (this is especially the case because it is networked and can significantly reduce the unit cost of moving information). But in Chapter 2, I also briefly discuss why the Internet, like most technologies, might do more for the rich world than for the poor world.

Chapter 3 takes the theory presented in the preceding chapter and applies it to the part of the world where the most information has been gathered about the impact of the Internet—the rich country club of the Organization for Economic Cooperation and Development (OECD). Even in these countries, it is difficult to say too much about the Internet per se, so the chapter begins with a discussion of the impact of information and communications technologies (telecommunications, computers, and software) on economic performance. The evidence is that the ICT industry is producing more powerful computers and software for the same amount of money, that companies across the rich world are investing huge amounts in this equipment in ways that increase output per worker, but at the same time, that there is limited evidence of a dramatic "pure technological" impact from use, as would be captured in rising total factor productivity statistics. That might be the case because some investments in ICTs carry very low returns, and I examine why. Looking at the Internet in particular, I conclude that e-commerce appears to have had a fairly marginal impact to date, and I discuss how widespread Internet access on workplace desktops might even be a drain on productivity. I conclude Chapter 3 by looking at what the evidence from OECD countries might mean for countries of the developing world, with their comparatively small ICT industries and limited institutional and educational capacities to take advantage of the numerous opportunities that ICT investments do provide.

Looking at the more limited evidence from the developing world itself, I suggest in Chapter 4 that the impact of ICT production is and will remain muted, compared to its significant impact in the United States. On the use side, I note the incredibly rapid rollout of ICT in poor countries since the late 1990s—suggesting the significant role that communications can play in fostering development and downplaying fears of a digital divide. At the same time, I note that advanced uses of the Internet, such as e-commerce, remain limited in the developing world, for a range of reasons connected with a generally low level of development. As suggested in Chapter 2, that raises concerns that the Internet could be a force for further divergence of incomes across countries.

Taking up this issue, Chapter 5 examines e-government in developing countries. There are some considerable opportunities for governments even in the poorest countries to improve transparency and efficiency through the Internet. At the same time, IT rollout is costly, and failures are frequent. Calls for widespread Internet access in schools in the developing world in particular appear to be divorced from the reality of education budgets. High costs and low benefits again suggest that government-run public Internet access programs may be a misallocation of scarce resources in the poorest countries.

In Chapters 2 through 5, I show there is strong evidence that businesses in developing countries already exploit the Internet where it has the highest value. Evidence for especially strong total factor productivity impacts is weak, however, and there is considerable support for the proposition that the broader environment in many developing countries may be unsuited to widespread use of advanced Internet applications. This evidence suggests that policies regarding the Internet in developing countries should be based on a "no regrets" strategy, rather than significant government intervention to favor the Internet over other investments. Such a policy is laid out in Chapter 6, which also discusses the role of donor countries. In terms of the impact that rich countries have on poor countries in the area of the Internet and development, aid policies matter less than these countries imagine; a halt to the attempt to strengthen intellectual property rights worldwide, for example, may foster the growth of the Internet in poor countries more than expenditures on unsustainable "pilot" telecenters. Chapter 7 concludes by comparing the social impact of the Internet worldwide to the economic impact discussed in previous chapters, finding there the same picture of mixed and muted effects.

Our past inability to predict the future impact of technologies, and perhaps in particular information and communications technologies, is the stuff of cliché. For example, in 1876, Alexander Graham Bell was touting his new invention (the telephone) throughout the United States and gave a presentation at the White House. There, President Rutherford Hayes turned to him and said, "That's an amazing invention, but who would ever want to use one?" A little less than 100 years later, Thomas Watson of IBM predicted a world market for five computers. But just as some have underestimated the potential of a technology, with even greater regularity, visionaries have forecast new worlds emerging as the result of a technological advance, forecasts that have proven universally overoptimistic. The death of paper has been predicted at least since a 1975 *Business Week* article, and the death of the pencil was announced by the *New York Times* in 1938.[24]

Looking at this latest information "revolution," Tom Peters, author of the best-selling business book *In Search of Excellence*, argues that today "we are in the opening innings of a world-class, unprecedented mess" created by the Internet, and because of this mess "we'll all get it wrong far more often than we get it right."[25] That seems a plausible line, based on our past record of prediction. It may be that this book will turn out to be excessively pessimistic in its estimates of impact and of potential. But the very fact that we are so bad at prognostication suggests the need for considerable caution. Investing billions of dollars in projects supporting Internet rollout, in environments where dollars are scarce enough that alternate uses with confirmed track records of reducing poverty are going short of funds, may be a high-risk strategy.

Notes

1. I was one of them.
2. Census Bureau 2003.
3. Some further examples: Freeman Dyson, professor emeritus of physics at the Institute for Advanced Study at Princeton, writes in *The Sun, the Genome and the Internet* that he "dreamed" that "solar energy and genetic engineering and the Internet will work together to create a socially just world, in which every Mexican village becomes as wealthy as Princeton." In *Cyborgs, Simians and Women: The Reinvention of Nature*, feminist Donna Haraway argues that the Internet would also end sexism (quoted in Cassidy 2002). Jeremy Rifkin declares that we are in "the age of access," with a whole new view of moral systems. For the people of the twenty-first century, "personal freedom has less to do with the right of possession and the ability to exclude others and more to do with the right to be included in webs of mutual relationships." That may all be true, but if so, someone had better tell my good friend Leo Moss, because despite being 100 percent twenty-first century (he is four), he nonetheless seems to be developing a strong sense of property rights regarding the sharing of toys with his brother.
4. Lewis 1999. Louis Rossetto, in his first column as editor of *Wired* magazine, upped the ante by suggesting that the Internet was the greatest invention since the discovery of fire (quoted in Cassidy 2002).
5. The Friedman quotes and ideas all come from his columns in the *New York Times,* www.nytimes.com, specifically, the columns printed on July 20, April 16, and

February 13, 2001; December 19, November 20, September 29, August 8, July 25, March 31, February 11, and February 1, 2000; December 8, March 9, February 2, and January 1, 1999.

6. Negroponte 1995.

7. November 25, 1997, on CNN, http://www.cnn.com/TECH/9711/25/internet .peace.reut/.

8. Similar hopes were expressed of earlier communications revolutions: the chairman of Cable and Wireless suggested world peace would be brought by the telegraph, and Marshall McLuhan was arguing in 1964 that the computer would usher in "a Pentecostal condition of universal understanding and unity" (quoted in Berman 1988). And Negroponte was hardly alone in his suggestion that national borders would be significantly weakened by the Net (see Franda 2002 for a review).

9. See http://www.digitaldividend.org/pdf/0501bw01.PDF. One person attending did raise some doubts about the utility of the Internet for poverty alleviation (see the conclusion of the book), but apart from the Luddite opposition of Bill Gates, the meeting was of the firm opinion that overcoming the digital divide was a global priority (http://seattlep-i.nwsource.com/business/gate19.shtml).

10. *Business Week*, August 28, 2000.

11. At the same time, activists on a nearby beach burned a laptop to protest the meeting's focus on IT rather than debt relief and the environment. The picture was e-mailed to listserves around the world dedicated to debt relief and the environment.

12. G8 2000.

13. Malloch Brown 2000.

14. If those measures did not appeal, they ranked themselves on "e-leadership"—how committed senior government officials were to e-based reform.

15. And, at a lower level of interest (and impact), Charles Kenny's work "Prioritizing Countries for Assistance to Overcome the Digital Divide" (Kenny 2001).

16. See http://www.bridges.org/ereadiness/tools.html. In the end, the main thing we learned from this profusion of studies was that rich countries had more Internet users, more IT graduates, and more government websites than did poor countries, but some poor countries did as well as rich ones on "e-leadership." A review by Maugis et al. (2003) concluded that "these studies are fraught with uncertainties and ambiguities in both theory and practice and lack robust foundations for empirical analysis. . . . As such, they provide little guidance for business and government."

17. Altig and Rupert 1999.

18. Weisburd, Lum, and Petrosino 2001.

19. "Sexual Harassment Legislation Victory for Women in India," available at http://www.apc.org.

20. Humphreys et al. 2003.

21. http://www1.worldbank.org/publicsector/egov/gyandootcs.htm.

22. Center for Electronic Governance, Indian Institute of Management 2002. See also Cecchini 2002.

23. See Forestier, Grace, and Kenny 2001 for a review.

24. Brown and Duguid 2000: 19. And technologies have been behind more than one attack of "irrational exuberance," or the madness of stock-buying crowds. The Internet bubble has its antecedents in the railway bubble. John Cassidy, a business journalist, notes that the slew of magazines that were launched to cover the Internet revolution bear uncanny resemblance to titles such as *Railway World, Railway Express, Railway Examiner*, and *Railway Review,* which were launched in 1840s Britain, only to fail after the financial crash of 1847. See Cassidy 2002.

25. "Source: 2001: A Business Odyssey with Tom Peters," *Georgia Tech Alumni Magazine*, Winter 2001. Given the record of the companies selected by Peters for their excellence, modesty in the predictive power of pundits is well taken (within two years of the publication of *Excellence,* the best-selling business book of all time, fourteen of the companies he had selected were floundering, according to *Business Week*). See Sherden 1998 for a discussion.

2

The Link Between Technology and Growth

The Record of Development Panaceas

The Internet is far from the first proposed silver bullet for underdevelopment. After World War II, theories based on physical capital investment were highlighted (based on the work of Evsey Domar in 1946), followed by human capital—health and education—in the 1970s. By the 1980s, the "Washington Consensus"—policy reform toward an open market economy—was seen as central. Since then, Nobel Prize–winner Douglas North has argued that underpinning economically successful regimes was a strong network of property rights and decentralized, democratic decisionmaking structures, a view he labeled *new institutional economics.*[1] More recently, followers of Harvard's Robert Putnam have argued that the reason developing countries were lagging behind was too little "social capital," which is, put simply, the strength of relations between individuals which encourage information transfer and exchange.[2]

In his book *The Elusive Quest for Growth,* Bill Easterly has done a wonderful job at skewering the silver bullet of investment. He notes that if the early theories linking investment to growth had worked as predicted, Zambia's GDP per capita would have risen from $900 in 1960 to $20,000 today—the real figure is closer to $600. The same margin of error that applies to investment may well also apply to more recent underdevelopment panaceas. Lant Pritchett of Harvard suggests that the cross-country evidence points to a weakly negative relationship between education stocks and growth.[3] Dani Rodrick, again at Harvard (apparently a hotbed of contrarian thinking on economic growth), has argued in a number of papers that the trade-growth relationship is also weak.[4]

Development silver bullets share a common property—a view of the world that is essentially a "view from nowhere," independent of time and location.

The whole world can be made better using a single set of rules about how people behave and how the world works. Especially regarding development panaceas, the evidence seems to point starkly in another direction. If and when the rules of any particular theory of development work depends on the environment in which they are tried. For example, Jeffrey Sachs of Columbia University has suggested that even if African countries followed all the policies suggested to them by believers in the Washington Consensus, the continent would still grow slowly because it is hampered by factors such as poor soils and physical isolation.[5]

It appears that the impact of policies, institutions, education, and investment on economic development depends on the circumstances into which they are introduced. It is likely that the same is true of technology, another entrant as a development cure-all. One early example of the technology quick fix was the campaign to increase tractor use on African farms in the 1960s, based on the idea that African farming was not modern enough, and a technological fix—mechanization—was the answer. Within a few years, most tractors were lying rusting and unused in the fields, defeated by a combination of lack of spare parts and fuel, as well as farming practices inappropriate to mechanization.[6]

But whatever our uncertainties about the causes of economic growth and the context-specificity of the impact of particular technologies, we do know that technology in the broadest sense of the word must have a major role to play—in part because the definition of technology as used by development economists is so broad. The rest of this chapter discusses the role of technology in growth, the importance of context in determining its impact, and what this might mean for the impact of technologies in general, and the impact of the Internet in particular, on economic growth in developing countries.

Defining Technology

Technology in the usual usage of the term conjures images of lab-coated technicians mixing chemicals, massive underground tunnels filled with atom-smashing equipment, hypersonic planes, or multi-terabit electronic pets that can be house-trained and taken for walks.[7] Development economists have a different image of technology. It would encompass the output of lab-coated technicians, but it would also include double-entry bookkeeping, total quality management, just-in-time delivery networks, labor regulations, unions, conglomerates, and political systems. It is, indeed, far simpler to define what technology is to a development economist by explaining what it is *not*. In short, technology is everything that affects the level of economic output, except capital or labor. As we shall see, this simple definition is in fact a powerful worm-can-opener, and so it is worth taking some time to understand this concept and what it means.

In most empirical discussions of economic growth, the impact of technology is measured through estimates of total factor productivity (TFP)—the efficiency with which given quantities of capital and labor are combined to increase value. One example of such a technology would be the steam engine in its first widespread usage as a mine pump. Prior to the steam engine, imagine that one worker and eight horses were required to pump out 100 gallons of water from the mine every hour. Then the mine owner sells the horses and, for the same price, buys a steam engine that, overseen by the same worker, can pump 200 gallons of water. To keep things simple, imagine the coal to run the engine costs the same as the feed for the horses, that the horses and the steam engine have the same expected working life, that the skills required to manage eight horses have a fair market value similar to those required to manage a steam engine, and no doubt more. But, in the end, one worker combined with eight horses' worth of capital has become twice as efficient thanks to the technology of the steam engine. In this case, growth in total factor productivity has increased output by 100 percent.

A second example would be Adam Smith's pin factory, highlighted in his discussion of the wealth of nations.[8] Here, the technology involved is the more efficient organization of labor into a factory. The same number of workers with a similar amount of equipment (capital) can produce a lot more pins in one factory than they could if they were "organized" in individual cottages. Smith does complicate the story by mentioning the skill of workers as well (what economists term "human capital"), but it is to an organizational—technological—change that he assigns most of the increased output. A single unskilled workman could "scarce, perhaps, with his utmost industry, make one pin in a day." Conversely, in a system where processes are separated: "One man draws out the wire, another straights it, a third cuts it, a fourth points it, a fifth grinds it at the top. . . . I have seen a small manufactory of this kind where ten men . . . could make among them upward of forty-eight thousand pins in a day."[9]

If, given the same input of supplies, equipment, and labor, a change allows you to produce more final goods, that change is technology. It could be a steam engine, a new chemical process, or a more efficient robot, but it could also be a new way of laying out the production process (like the pin factory) or a new management system. At the national level, it might be a change to a different political or social model or the appointment of a smarter chairman of the Federal Reserve or a president or Congress who aren't beholden to special interests.

Again, the technology that drives TFP is *very* broadly defined. Indeed, TFP increases are calculated as a residual, to account for "everything else that impacts growth" after looking at the contribution of capital and labor. Growth accounting procedures measure increases in stocks of physical capital (the quantity of assets such as factories and roads, measured by their dollar value), human capital (the education stock, usually measured by average years of schooling in the working population), and labor (measured in numbers of people of working age). The procedures then estimate how much output growth can be "expected" from such increases in capital and labor stocks. Total factor productivity is defined as the actual measured growth of output minus the growth rate expected from increases in capital and labor stocks. As a measure of the efficiency with which capital and labor combine to produce output, TFP is calculated as the fudge factor needed to balance the equation from this "growth accounting" exercise.

As such, technology as defined by growth accountants includes "business technology" (management techniques and systems), "political technology" (forms of government and institutions), and "social technology" (modes of human interaction)—once more, it includes everything that might affect output, except physical and sometimes educational capital or labor.

The Role of Technology in Growth

With such a broad definition of technology, it is perhaps not surprising that it is important to the growth process, even if it is rather difficult to pin down *how* important. Technology as defined by TFP is open to a wide range of assumptions about the importance of various factors to output that can dramatically alter its aggregate size. As such, calculating the scale of TFP growth is a process laden with more or less arbitrary decisions that can dramatically alter results. Depending on how we define physical capital, labor stock, or human capital; how we measure the growth of stocks of capital; what, if any, allowances we make for technological change "embodied" in capital investment or market failures such as scale effects; whether we believe that capital can be substituted for labor one-for-one; what share we give to human over physical capital in output estimates; and a range of other more-or-less arbitrary assumptions, we can come up with markedly different estimates for the residual from a growth accounting equation that is defined as total factor productivity. For example, one recent study looked at the impact of changing just two of these assumptions within reasonable bounds on long-term TFP growth in Korea and produced estimates of annual TFP growth between 1960 and 1997 that ranged from 3.0 percent of GDP to minus 1.4 percent.[10] In other words, technology was either the driving force behind economic growth in Korea during its "miracle" period, or technological change as measured by TFP actually went rapidly *backward* over that period in Korea, all depending on the (reasonable) assumptions made.

Because the assumptions made by economists can significantly influence estimates of technology's impact on growth, the extent to which technology has been important in practice in regions like East Asia remains hotly debated. Depending on the exercise one chooses to believe, the East Asian growth miracle can be explained almost entirely in terms of high rates of physical and human capital accumulation (by investment in goods and education) or largely on improvements in the productivity of human and physical capital due to technology adaptation and advance.[11] Even using similar methodologies (weights and procedures), the measured contribution of factors varies greatly by country and period.[12] Economic heavyweights, including Paul Krugman of MIT, spilled ink like blood in *Foreign Policy* and the *Journal of Economic Perspectives* over the role of capital versus investment in explaining the glory days of economic growth in East Asia. Jesus Felipe of the Asian Development Bank concludes his survey of the exchange by noting "this work has become a war of figures. . . . It seems that [by] re-working the data one can show almost anything."[13]

Richard Nelson, a professor at Columbia University, suggests the entire effort to calculate TFP may be flawed for these reasons:

> I am very uncomfortable with the attempts of my colleagues in economics to "divide up" credit for economic growth between capital formation, education, and other input increases, and technological advance. I think it important to understand the economic growth process as involving strong interaction among various elements. . . . It makes little sense to ask how much growth we would have experienced if capital per worker had increased as much as it has and we had no technological advance, because it is technological advance that enabled the growth in capital intensity that we have experienced.[14]

Nelson makes perhaps three central points: (1) we don't know how important technology is for growth, but (2) we do know it must play a big role, (3) a role linked to changes in the stock capital and labor. And we know that technology is important in part because the other competitors are so weak. As Bill Easterly and Lant Pritchett point out, it isn't investment alone (or Zambia would be rich) and it isn't education alone (or the whole developing world would be). Under the growth accounting framework, pretty much the only thing left is technology. And so, in Joel Mokyr's elegant phrase, technology must be "the lever of riches."

The Nature of Technology and the Divergence in Global Incomes

That technology writ large may be the primary source behind economic advance suggests poorer developing countries have a number of reasons for concern. First, "stocks" of technology seem to be highly concentrated in particular parts of the world. The average Bangladeshi lives on less than a dollar a day, whereas the average American lives on close to 100 times that. If technology is what drives differences in income, then there must be much more technology in the United States than in Bangladesh.

It should also be noted that if new additions to the global stock of technology are the major factor behind growth, technology must be *consistently* more favorable to wealthier countries over the long term. By and large, rich countries today were comparatively richer in the past, and over the long term, we have seen a consistent pattern of divergence in income between the rich and poor countries—what Lant Pritchett labels "divergence, big time." There have been some comparatively large winners over the last 100 years in the growth race between the countries that were already relatively rich in 1900, including Italy, Norway, and Japan. There have been some comparably slow growers in this group—Argentina, New Zealand, and the UK, for example. But as a rule, rich countries have stayed rich. Even the long-term laggards just mentioned still manage to scrape by, with Argentina—the poorest of the three—still at the upper end of the upper-middle income bracket. Between 1870 and 1989, two-thirds of the present high-income industrialized countries averaged annual GDP/per capita growth rates within 0.2 percent of the US rate.[15] And poor countries remain poor, with the African continent the usual example. Worse, the gap between the rich and poor is yawning ever wider. Long-term growth data for forty-two countries, compiled by Angus Maddison at the OECD, indicate that the average per capita income of the poorest ten of those countries in 1900 was $637. The richest ten countries in 1900 had an average income of $3,696, approximately six times as much as the poorest. In 1992, the poorest ten countries had an average income of $1,831, compared to the richest ten's average of $18,765—approximately ten times as much as the poorest.

That rich countries stay relatively rich and get relatively richer and that different stocks of technology may help explain it would come as a surprise to early development economists. They used a model that suggested that the technology stock was the same the world over. What is now called "exogenous growth theory," which dates back to Robert Solow's 1956 model regarding investment and growth, predicted that the long-run rate of growth worldwide was determined primarily by an expanding and globally available stock of technology. Their concept of technology was one of concepts themselves—free-floating knowledge that, once created, was available to all.

Instead, it must be that the technology that is important to growth is "sticky" and hard to move around. An exogenous theory that suggested otherwise began losing popularity by the mid-1970s precisely because it became increasingly clear that countries weren't all growing, so technology couldn't be free-floating. Indeed, poor countries were falling ever further behind, suggesting technology must be very sticky indeed.

Nonetheless, technology remained central in the set of theories that replaced exogenous models—"endogenous" growth theories of the type put forward by Paul Romer of Stanford University. The new theories argued that technology played a similar role in promoting growth as suggested by exogenous theories, but that the rate of technological change in countries was not the same worldwide, instead being determined by factors such as the investment rate, policy environment, or level of trade. Romer argued in a 1983 article that investment was a cause of economic growth, not just because more factories, infrastructure, and machines themselves produced more output, but also because more investment in those factories, infrastructure, and machines was linked to a faster increase in the technology stock available to a country. In turn, that allowed those same investments to produce more output. It was this growing stock of technology that drove increases in income.

The Theory of Embodied Technologies

Early endogenous growth theories such as Romer's contain the idea of a technology being "embodied" in human and physical capital (an idea also suggested by Richard Nelson's comments about the futility of TFP calculations, quoted above). For example, in order to benefit from the greater pumping power of a steam engine, you have to buy a steam engine and train someone to use it. In Adam Smith's pin factory, there were also clear linkages between the "technology" of efficiently divided labor and the need for human and physical capital: "a workman not educated to this business (which the division of labor has rendered a distinct trade), nor acquainted with the use of the machinery employed in it (to the invention of which the same division of labor has probably given occasion), could scarce, perhaps, with his utmost industry, make one pin in a day."

If investment in human and physical capital is necessary for the successful utilization of a number of technologies, that in turn might account for the phenomenon of divergence in incomes across countries. The technologies invented may be designed to increase productivity in countries with a particular mix of capital and labor. Inventors in labor-rich, capital-poor countries would be likely to look for capital-saving technologies that require comparatively large amounts of labor, whereas inventors in countries with shortages of labor might look for technologies that require considerable capital but few people to operate. If inventions occur in countries with a particular set of labor and capital endowments, those inventions will be of less use in an environment where relative supplies of labor and capital are different.

Because creating new technologies is a complex and frequently expensive process, the developed world creates far more technology than the developing world, as we'll see in later chapters. All this technology is likely to be designed with developed countries in mind—they are the larger markets. Furthermore, they are distinguished not only by their higher output but also by their greater stocks of and access to capital.

Again, theories of embodied technologies suggest the need for economies to have a particular mix of physical capital, human capital (skills), and labor in order to benefit from the potential productivity impact of a particular technology.[16] Assuming that most technology development occurs in the developed world to spur productivity in a capital-rich setting, most new technologies are likely to be both "embodied" in a significant quantity of capital and designed to save labor—the scarce resource in rich countries—and thus would be inappropriate for an LDC setting, where capital is rare and labor plentiful. Such a way of looking at technology and growth can produce worrying results for LDCs. Rich countries develop technologies that cannot be fully exploited in poor countries. They adopt these technologies, poor countries do not, and the income gap widens.

An example of the two types of technology that favor rich countries and poor ones might be the car and literacy, respectively. A motor vehicle is a piece of physical capital that embodies a good deal of technology (not least the internal combustion engine). A literate person is a "piece" of human capital that embodies technology, too—the alphabet, numbers, and punctuation. The technology behind literacy, however, requires less capital for embodiment than does the technology behind a motor vehicle. Poorer countries can much more afford to provide each person with the technology of literacy (by utilizing their abundant labor to make them literate) than they can to provide each person with the technology of the internal combustion engine (by using scarce physical capital to provide a car or a bus network).

Given that, it is not surprising that over time, the average number of cars per capita in a country at a given income level has changed little, and few countries have "many more" vehicles than their income level would suggest. Conversely, the average number of "literate people per capita" that a country at a given level of income can afford has improved markedly, and a number of countries have many more literates per capita than their income might suggest.[17] The technology of literacy is far more exploitable for LDCs than the technology of the car, but (at least from a theoretical standpoint) we would expect developed countries to produce more technologies with the features of a car than with the features of literacy.

The type of technology that is important for economic growth might be embodied in physical and/or human capital, the skills and machinery of the economy, then. Because labor-scarce countries are likely to be wealthier, to produce the predominance of new technologies, and to invent technologies that will be focused on saving labor, those characteristics might lead to significant income divergence across countries.

Beyond Embodied Technologies: The Role of Institutions

But if "embodiment" was the only explanation for divergence, we'd expect the world to look very different from the way it actually does. For a start, there would be a far stronger relationship between growing stocks of physical and human capital (investment and education) and economic growth—rather than the awfully weak one that we have seen. Secondly, no one in developing countries would bother with capital-intensive technologies such as a car, which is inappropriate for their capital-scarce environment. Yet people in developing countries—or at least some of them—*do* have cars. They have enough cars that traffic jams in cities like Bangkok and Lagos are legendary. Even Kabul has hideous traffic jams nowadays. More generally, allowing for output, people in LDCs have about as many cars as you'd expect them to have (in other words, about the same share of their income goes to buying cars as it does in rich countries). That's less than the number of literates they have but still suggests that the car is useful in developing country contexts. The same is true of a raft of other technologies. Given their income level, developing countries have as much in the way of stocks of capital that embody technology as you would expect.

If investors in developing countries are acting at all rationally, they only buy this embodied technology if they think it can make a return. Some of those investors may be wrong in that assumption. Indeed, the thesis of this book is that, with the case of the Internet in particular, some government investment decisions regarding computer and communications purchases may well have a low payout, just as there was a low payout to investing in Internet stocks in the United States in 2000. But it is highly unlikely *all* technology investors in developing countries are wrong. And, if that's the case, there are at least some productive uses of technology that require significant quantities of capital in developing countries. Again, that's a point that will reappear again in this book with reference to the Internet—*some* uses of the Internet (and some investments) really make sense in developing countries.

If a heavy capital endowment does not make a technology (entirely) inappropriate for a developing country, we would not necessarily expect new technologies to be, on average, a force for divergence. We are back to square one. It appears that (capital) rich and (capital) poor countries can both benefit from inventions designed in capital-rich settings.

Perhaps something else, or at least another form of technology, is also important for growth and divergence. Because technology as defined by TFP equations is very broad, including far more than technology embodied in physical capital, there is space for other types of technology to be more important for sustained growth than "formally" invented, embodied (physical), technologies (such as the Internet). Indeed, Paul Romer, the father of technology-driven growth theory, now argues that Wal-Mart's management of inventory data is probably a more significant source of economic growth than inventions such as the transistor.[18] Romer says that physical technologies that are traditionally researched and developed cannot be a central cause of the process of economic growth, and one reason to believe him is that the number of scientists and engineers employed in R&D in the United States increased fivefold from 1950 to 1990, while US growth rates fell over that period.[19] Conversely, there is a growing body of literature connecting institutional structures (processes and rules) in areas such as finance, law, and government with long-run economic growth across countries.[20]

But again, the idea that "nonphysical technologies" like processes, rules, and ways of doing things matter more to overall economic growth does not bode well for developing countries. Institutions are notoriously difficult to transfer across countries. "Technical assistance," the bit of development aid aimed at that very transfer, is routinely dismissed as one of the least successful parts of aid programs.[21] And it is fairly easy to imagine why. A car can work pretty much as well in Lagos as it can in L.A. Wal-Mart's inventory management system really wouldn't. Nature is reasonably similar worldwide, and so technology that assists human mastery over nature works worldwide. But institutions are really very different, and so a person's mastery over social systems is very specific to time and place. Again, it is easy to be a successful mechanic in Detroit or Dakar, but it is not so easy to be a successful inventory specialist. Of course, if it is hard to transfer institutional technology, and if it is behind growth, then that is bad news for LDCs.

Even if (and it only seems probable) "invented" technologies share the stage with institutions as sources of growth, it may be that they interact so as to keep poor countries poor. A range of endogenous theories, in particular, suggest that the extent to which a particular technology *actually will* (rather than potentially might) have an impact on growth does depend on a range of environmental factors beyond the relative scarcity of capital: perhaps social mores, trade policies, the presence or absence of a central bank, or a country's status as democracy or dictatorship might play a role. If these factors are in turn linked to a general level of development, that might also explain a divergence result generated by differential abilities of rich and poor countries to benefit from new technology.

Jared Diamond, in *Guns, Germs, and Steel,* points out that the history of invention suggests that the use, if any, to which a new technology is put does indeed depend very much on the nature of a society into which it emerges. Ancient native Mexicans never used the wheel for anything but toys because they had no suitable animals to draw carts, the Japanese continue to use the (cumbersome) kanji in addition to the (far simpler) kana alphabet because of prestige, and QWERTY dominates global keyboards because everyone (except this author) has already learned how to type on them. The "nature of a society," in economic terms, is in fact a set of technologies itself. How great an impact a technology has, then, depends crucially on what technology already exists—and, in particular, may depend on the institutions already in place.

Implications for the Role of the Internet in Development

That technology change and adoption are potentially the major sources of economic growth is clearly good news for those who would see the Internet as a globally transformative invention. And indeed, the Internet has features that suggest it may be a particularly powerful technology for economic growth.

First, through the benefit of "network externalities," investment in ICTs by one company or individual spills over, providing benefits to others connected to the network. If I have an AOL account and no one else does, the number of times I will hear "you've got mail" in any one day is likely to be limited by my capacity for electronic onanism. But for each person who joins the network, my computer is increasingly likely to chirrup. If my wife, brother, and parents join, for example, I now have four people to exchange e-mails with. The benefit of the network to me has increased, while its cost (to me) remains the same.

Some IT firms—most particularly Microsoft—have used network externalities to generate huge profits. Listen to any disgruntled inductee in the cult of Apple, and they will willingly tell you at great length that the only reason everyone uses Microsoft products rather than the obviously (to them) superior Apple versions is because everyone uses Microsoft products. To put it another way, because it is difficult to transfer documents or spreadsheets or (even more) programs from a Windows-operating computer to an Apple, anyone who wants to communicate with Windows operators will choose a Microsoft-equipped computer over the Apple. If my brother and parents and wife have Windows, I'll probably get Windows, too. Apple recognized this problem early on, and it is no doubt one factor behind their decision to heavily discount products for students, to try and build up a critical mass of users before Microsoft got there. But Bill Gates did better, putting his software on millions of IBM clones. It might, so easily, have played out the other way.

Regardless, the ability of Microsoft to accrue billions in profits through the exploitation of network externalities shows the potential power of similar externalities linked to the Internet to generate broader economic impacts. And that is especially the case regarding what its proponents see as a general purpose technology on a par with electric lights and motors, technologies with applications across the economy, as opposed to a better chainsaw design, for example, that could only revolutionize logging. (Given the broad definition of technology and thus the small potential impact of the average new technology, only the exceptional technology with wide applications will have a noticeable impact on macroeconomic indicators.)

A second reason the Internet may produce especially high economic returns is that it can dramatically reduce transactions costs—the costs of exchanging goods and services in a market or within an organization. Lower transactions costs lead to the creation of new markets and more efficient operation of existing markets not accounted for in the initial decision to purchase the system. The Internet is a particularly powerful tool for the transfer of information goods—software, news, Beanie Baby catalogues—that have peculiar economic characteristics, in that they are expensive at first to produce but very cheap to reproduce (especially with the Internet). Such goods themselves have the potential to reap higher returns, the further they spread.

But the general discussion about the role of technology in growth also suggests a number of potentially worrying implications for the Internet's role in development. First, the Internet, as an "invented" technology, may actually be less important to the growth process worldwide than institutional technologies.

On top of that, if the Internet is a "standard" technology, it is likely that its effects will be far larger in the already wealthy world than in the developing world. Proponents often see the Internet as a "leapfrogging" technology, which will have a greater impact in the developing world than the developed. It would be an unusual technology if that were the case, and we will see in later chapters that the features of the Internet suggest it is very unlikely to be unusual in that way: the Internet is a human capital– and physical capital–intensive technology. In addition, we will see that the very feature proponents put forward as a reason to believe the Internet is a potentially revolutionary technology—its networked nature—may actually force people and companies in developing countries into suboptimal investment decisions involving this capital-intensive, less than suitable technology.

Perhaps of even greater concern, developing countries are unlikely to have the correct institutional environment to benefit from the Internet, and those institutional factors may be the most important in determining the potential impact of a technology. At the broadest level of abstraction, as we have seen, the theoretical impact of the Internet is part of a model of economic growth based on the accumulation of technology and ideas rather than one based on factor accumulation (human and physical capital). But that characteristic poses significant problems for LDCs, where the institutions required to operate in such an economy are at their weakest.

Policy Implications

Technology is important for generating wealth and is a source of the divergence between rich and poor countries in terms of income growth. That divergence may occur in part because the successful exploitation of the average "lab coat" technology involves capital endowments—or other features of an economy—that are present in rich countries but absent in poor ones. The divergence may also occur because the type of institutional technologies that drive growth are even more specific to location than physical technologies.

Different models of the role of technology in development suggest markedly different policy conclusions about the role of governments in fostering technology development itself. The analysis above suggests that different institutions and capital endowments may dramatically alter the returns to adopting a particular technology, further suggesting the absence of a "one-size-fits-all" policy package regarding responses to a new technology.

It is worth pointing out here that the basic justification for any government intervention rests on the idea that the market, acting alone, will produce suboptimal results—and in the case of growth-promoting technologies, there is a strong case to be made that markets alone will underinvest. To a significant extent, physical capital acts like a "proper" market good. It is excludable (I can stop you using my factory by locking the door) and rival (we can't both use the factory at the same time). Ideas and technology are excludable and rival to a lesser extent. We can both use them at the same time, and (absent patent protections and the like) I cannot stop you using them. Perhaps the central insight of growth theory is that this strange nature of technology means that growth might be predominantly connected with a "market failure." But in turn, that increases the importance of having government institutions (education and research, intellectual property rights, and the courts) to manage the market failure. A belief in the centrality of technology to growth may well suggest an *increased* role for government in the future.

But whether there is a role for government in fostering the development of a particular technological application or technology-based sector depends, of course, on the application or sector being especially growth-promoting. Total factor productivity growth provides some indication of such a technology at work, because it suggests higher returns on investments than would usually be expected. And so, if there is evidence that the Internet (or the ICT sector more broadly) can promote TFP gains, it would constitute some support in favor of a government role to promote use of and access to the Internet and ICTs more broadly.

In Chapter 3, I discuss evidence for TFP gains from the ICT and Internet sectors where they should be easiest to find—in developed markets where there are larger stocks of ICT and more data to capture its impact. That allows me to evaluate how much ICT has been a particularly powerful lever of riches in wealthy countries. The discussion focuses around TFP gains and explores why they do not appear to be particularly widespread across economies. I look at potential negative net externalities as one possible cause. I then look at the Internet in particular to evaluate the potential impact that it has had on productivity and economic growth. I also discuss, based on empirical evidence from wealthy countries, whether the Internet is likely to be a pro-poor technology, suggesting even greater TFP gains from use in developing countries. Is the Internet like most technologies in that it is likely to have a larger impact in wealthy countries, or is it that rare technology that is better suited to a developing country environment? That discussion will consider what is necessary, in terms of institutional capacity and physical and human capital, for the successful exploitation of the new technology.

Notes

1. North 1992.
2. Putnam 1993. In a related strain, Francis Fukuyama has suggested the end of history (or global development towards wealthy liberal market democracies) is only being delayed by a lack of trust in developing countries (in which category he includes France).
3. Pritchett 1996b.
4. See, for example, Rodriguez and Rodrik 1999.
5. Sachs and Warner 1997. See Kenny and Williams 2001, for a longer discussion.
6. Wade 2001.
7. For auditors in the Government Accountability Office, it might also conjure images of huge cost overruns.
8. Smith waxed lyrical on the subject of the pin factory, so much so that the *Wealth of Nations* was nearly called *Just Sew Stories*.
9. Smith 1976.
10. The two were assumptions about returns to scale and shares of physical and human capital in output. See Ghosh and Kraay 2000.
11. See, for example, Young 1995 and Crafts 1998 (on the side of capital) versus Klenow and Rodriguez-Clare 1997 (on the side of total factor productivity).
12. Crafts 1999.
13. Felipe 1999.
14. Nelson 1998.
15. Pritchett 1997.
16. See the collection of articles in the *Quarterly Journal of Economics* 113, no. 4 (November 1998).
17. See Kenny 2005.
18. Perkins and Perkins 1999b.
19. Keely and Quah 1998.
20. See, for example, Acemoglu, Johnson, and Robinson 2001.
21. See, for example, Cassen et al. (1994) on the mixed record of technical assistance, especially in Africa.

3

Information and Communications Technology (ICT) in the Industrialized World

All fixed, fast-frozen relations, with their train of ancient and venerable prejudices and opinions, are swept away. . . . All that is solid melts into air.[1]

KARL MARX WAS TALKING about the final stages of capitalism in this quote. It happens to describe quite well what went on in the US stock market at the end of the 1990s, as price earnings ratios became anciently opinionated (it was hard to measure the ratio when a company had no earnings) and valuations based on anything more than measurements of the bubble became venerably prejudiced. At the end of the day of its 1999 initial public offering (IPO), for example, Priceline.com was valued at almost $10 billion: the year before, it had sold just $35 million worth of tickets and had generated losses of more than $114 million ($1.5 million of which came from selling tickets for less than the company had paid to buy them).[2]

Behind those truly impressive stock numbers lay a belief that the US economy was entering a different era (air over solid), led by dot-coms that were the ultimate in weightless businesses. To some extent, that belief was correct, but in this chapter I suggest that the scale of the transformation—in the United States, but even more so in the OECD as a whole—was not revolutionary, nor was the nature of the transformation one that should necessarily bring comfort to developing countries in their search for so-called growth poles.

Because of their close interrelationship at the macroeconomic level, it is very difficult to separate the impact of the Internet from the broader impact of information and communications technology investment on the economic performance of the US or OECD economy. I begin this chapter with a discussion of that broader impact in the 1990s, before turning to what we *can* say about the Internet, an analysis of the nature of the impact, and finally, possible implications for LDCs.

ICT, Labor Productivity, and Total Factor Productivity in the United States

Something big went on in the US economy in the 1990s. The Internet, although still a novelty, became ubiquitous. Never has a communications tool reached greater than 50 percent household penetration so fast. The hardware, software, and communications industries increased their share of GDP from 2.7 to 4.2 percent from 1990 to 1998, and information and communications investment as a share of total investment reached 25 percent by the end of the century (up from closer to 10 percent two decades before).[3] By the end of the decade, more than 13 million Americans held IT-related jobs, and the rate of growth in these jobs over the 1990s was six times as fast as overall job growth.[4] The number of jobs requiring computer skills increased from 25 percent of the total in 1983 to 47 percent in 1993. By 2000, about 60 percent of the nation's jobs demanded computing skills—and paid an average of 10 to 15 percent more than jobs involving no computer work.[5] The Internet played a significant role in this transition, driving up demand for telecommunications and IT hardware and software as well as the skills required to use them.

All this investment and expansion had a significant impact on US growth rates. Nobel Prize–winning economist Robert Solow of MIT must heartily regret making the snappy observation in the late 1980s that you could see computers everywhere but in the productivity statistics, because a bevy of economists studying ICTs never tire of quoting it at the start of papers that go on to suggest just such a link.[6] Not least, it is likely that the growth in investment in ICTs in the United States from average levels between 1979 and 1995 to those between 1995 and 2000 was responsible for as much as 0.4 percent additional annual growth in labor productivity—one-third of the total increase in labor productivity between those two periods.

Nonetheless, Solow's observation remains potentially correct—depending on what one means by "productivity." In fact, Solow meant *labor* productivity (output per worker), and he probably couldn't find a link looking at data from the 1980s because computer investments were such a small percentage of total investment. During the 1990s, investment in ICTs rose rapidly, and it is widely agreed that we saw the results in the labor productivity data.[7] But if Solow had made his observation regarding *total factor* productivity (the efficiency with which labor and capital are combined to produce output), the jury would still be out, at least regarding investment in ICTs.

The productivity story of the late 1990s and early 2000s, based on revised data from the US government, is now accepted by both skeptics such as Robert Gordon, an economics professor at Northwestern University, as well as optimists such as Steve Oliner and Dan Sichel of the Federal Reserve.[8] First, the United States grew faster in the late 1990s than it had in years previously (with labor productivity increasing at 2.5 percent from 1995 to 2000). And labor productivity spurted even farther ahead in the first three years of the twenty-first century (up to 3.4 percent).

Second, the two major sources of that labor productivity growth were ICT industries and ICT investment. Above-trend TFP growth in ICT-producing industries meant that computer companies were able to produce smaller, better, faster machines (and software and cables) using the same amount of capital and labor to make them. At the same time, corporations spent $1.1 trillion on computer hardware between 1990 and 1996, and at that point investment began to show up in the growth statistics.[9] Just assuming normal returns on all that investment in information and communications suggests a significant impact. Companies were buying increasing numbers of computers (and software packages), and it is likely that employees were producing more goods and services by using them.

Between 1995 and 2002, these two factors (increased TFP growth in the IT-producing sector and increased investment in ICTs in the rest of the economy, or "ICT capital deepening") accounted for as much as an annual 0.76 percent increase in labor productivity growth over the period prior to 1995. That is an impressive performance that should not be downplayed. An extremely rapid growth in the capital stock of computers and software, combined with more efficient IT production, led to a considerable increase in output per worker.

A harder question to answer is whether investment in and use of ICTs produced "supra-normal" returns (reflected in dramatic increases in total factor productivity) of the type that would suggest the technological advances in the sector could be a significant and sustainable source of more rapid economic growth over the long term. Sadly, what evidence we have on this is not reassuring.

When we say that ICT investments promote growth through "capital deepening," it is, of course, only an estimate based on the growth accounting formula. A certain amount of investment is meant to lead, if it generates "normal" returns, to a certain amount of growth. As Bill Easterly's example of Zambian investment discussed in Chapter 2 (a lot of investment, negative growth) suggests, these estimates can be unreliable. Having said that, if there were very low returns on investments in ICTs in particular sectors within the United States, we'd expect to see that reflected in lower TFP growth—because TFP is the residual in the growth accounting formula, which assumes a certain return on capital. Again, TFP measures the difference between actual output and that output which would be expected, given a known level of capital and labor input and "normal" returns on those inputs. A big (positive) difference, with actual output far higher than expected output, is high TFP growth. You can see it also as high returns on labor and capital inputs. Conversely, low TFP growth can be interpreted as reflecting lower than expected returns on investments or labor inputs.

So, if we see high TFP growth in sectors of the economy that invested relatively large amounts in ICTs, we can interpret that as high returns on ICT investments. And if we see low TFP growth in sectors that invested relatively large amounts in ICTs, we can interpret that as comparatively low returns on ICT investments.

There are cases in which ICT-investing industries also saw high investment returns reflected in TFP growth. Robert Gordon notes that wholesale and retail traders, for example, saw significant productivity gains and also significant investments in ICT. However, the broader evidence points to the idea that something else was at work than high returns on ICTs themselves. All the retail productivity growth can be attributed to new establishments replacing less productive stores. The average existing establishment saw zero productivity growth. That suggests that the driving force behind productivity gains was not ICTs per se (although they might have a part to play in the story) but Wal-Marts and Best Buys, with their economies of scale and efficient national logistics operations, replacing mom-and-pop corner stores. (The lack of Wal-Marts in Europe might also help to explain why the rest of the OECD failed to see the productivity gain that the United States did, as we shall see.)[10]

McKinsey and Company's report, *US Productivity Growth, 1995–2000*, echoes these results at a more general level.[11] It found that retail, wholesale, securities, telecom, semiconductors, and computer manufacturing, which accounted for 30 percent of economic output, accounted for nearly all the growth in labor productivity over the last half of the 1990s. These sectors, on average, were more IT-intensive than the rest of the economy. But the rest of the economy (the other 70 percent), while accounting for 61 percent of the US acceleration in IT intensity, managed only 0.3 percent labor productivity growth—and actually saw TFP growth of –0.3 percent from 1995 to 1999, compared to 0.4 percent from 1987 to 1995.

At the same time, within those six sectors that saw fast labor productivity growth (retail, wholesale, securities, and three ICT hardware sectors), a number of other factors (competition, regulatory change, cyclical demand factors, other innovations) were probably more important than IT investment itself. For example, according to McKinsey as well as Robert Gordon, Wal-Mart's 48 percent productivity advantage over its competitors in the mid-1990s, and productivity responses by competitors were based on factors such as economies of scale in warehouse logistics and purchasing and large-scale format, which were related to, but far from the same as, IT innovation. Overall, the McKinsey study concludes: "In rare cases, IT (including communications equipment) can deliver truly extraordinary productivity improvements. . . . In most cases, however, IT is just one of many tools that creative managers use to redesign core business processes, products or services."

Martin Baily, former President Bill Clinton's last chairman of the Council of Economic Advisers, also suggests that the evidence is "quite mixed" that US industries that saw faster growth in IT capital intensity saw faster labor productivity growth after 1995 (let alone TFP growth). He concludes: "In some cases, investments in information technology yielded little payoff. . . . In other cases, information technology contributed substantially."[12]

Opinions differ. Bill Lehr and Frank Lichtenberg, two economists at Columbia, claim there have been excess returns on investments in computers across a range of industries, compared to other types of capital (in the United States), especially in the presence of skilled labor.[13] However, it appears that, as with the Wal-Mart case, that may be the result of conflating ICT investment and other factors. Again, "excess" microlevel returns seem to disappear when other factors, including company reorganization, are taken into account.[14]

Further suggesting the lack of a positive link between ICT investment and total factor productivity performance are data showing that OECD countries and US regions that invested more heavily in ICTs did not on average grow faster—indeed, there are some data to suggest the reverse. The evidence from a recent study of OECD countries by Paul Schreyer is that there, as in the United States, investment in ICTs has jumped. The ICT share of nominal productive capital stock increased from 2.4 to 3.2 percent in France, from 1.2 to 2.3 in Japan, and from 3.6 to 5.2 in the UK from 1985 to 1996 (compared to growth from 6.2 to 7.4 percent in the United States). However, no evidence exists for the increase in TFP growth from 1985 to 1996 that would be expected in OECD countries if there were spillovers from ICT investment.[15] The UK, with the second-largest share of ICT in total capital stock behind the United States, has actually seen a TFP *slowdown* in the last few years.[16] The cross-OECD result is echoed by evidence from the United States at the state level, which finds that there was no increase in productivity growth in US states that were not intensively involved in the *production* (as opposed to use) of IT goods and services.[17]

Low Returns on ICT Investment in the United States

That the McKinsey report and Schreyer's work note some evidence of slowed total factor productivity growth across large parts of the United States and OECD economies suggests the possibility of an even more worrying conclusion—that the average computer investment has in fact underperformed. Indeed, many sectors that invested heavily in ICTs actually saw lower TFP growth, on average, than sectors that invested relatively little in ICTs. According to one study from the late 1990s, TFP change in the United States averaged around 4 percent in mining over the 1987–1997 period, compared to close to –4 percent per year in the banking sector—this despite the fact that IT spending as a percentage of output was highest in banking among the industries covered and second lowest (after construction) in mining.[18] Across industries, gross product per worker in IT-using goods and services fell an average of 0.1 percent from 1990 to 1997, compared to a 1.1 percent rise in non–IT intensive industries.[19] Although more recent data suggest that IT-using industries are performing better (perhaps because of a lag between investment and return in ICTs), the evidence against supra-normal returns does seem to remain. [20] It all suggests that the Solow paradox—widespread evidence of computer use and little evidence of (widespread) productivity growth—continues, at least regarding total factor productivity.

If it is the case that companies invested in ICTs that generated low returns, it may be that irrational exuberance in the stock market was only the most frothy manifestation of a bigger investment bubble that spread across the economy as a whole. It is clear, for example, that companies such as World Com overinvested in information and communications infrastructure. Robert Gordon notes that, in 2002, only 3 percent of intercity fiber-optic capacity was utilized, and the underutilization had wiped $2 trillion off the stock values of involved companies.[21] But it may also be that a whole range of other companies overinvested, as well—banks, auto manufacturers, booksellers. Again, the bulk of US companies' very considerable investments in IT produced healthy returns, but it is possible that the average IT investment did a little worse than investment in other capital, perhaps dragged down by a number of investments that produced zero or negative returns.

Irrational exuberance might have blinded companies to the full costs of ICT investment. One estimate is that each computer carries with it $5,000 in "stealth spending," including factors such as peer support and delays caused by crashing computers. Other, similar, estimates suggest that capital expenditures constitute only 20 percent of the cost of a computer to an organization. In companies using more advanced computing equipment, the capital costs are far higher than the depreciated cost of a $700 desktop. Microsoft was spending about $16,000 per year on each of its workstations on maintenance and operation during the 1990s, for example.[22] It may be that managers who were neophytes in the world of IT underestimated these hidden costs and so overinvested.

Certainly, the evidence on the success of IT projects might support such a view. The Standish Group's 1994 survey of IT executive managers found that only 16 percent of IT projects were completed on time, on budget, and with all features and functions as originally specified. About half (53 percent) of the projects reported by the IT managers saw budget and time overruns or a reduction in the original goals of the project. Budget overruns averaged 89 percent of estimated costs, time overruns on average more than doubled the time taken to implement the project, and reductions of functions and features averaged 39 percent of originally planned features in the initial design specifications. The remaining 31 percent of projects reported by the surveyed IT managers (beyond the 16 percent of successes and 53 percent of problematic projects) were simply canceled.[23] Another recent survey in the United States suggests that 42 percent of corporate IT projects are abandoned before completion, and 50 percent fail to meet CEO expectations.[24]

Economic Mismeasurement, Declining Returns, and Baroque Technologies

Another related possibility for apparently low TFP returns on ICT investment in the late 1990s is simple economic mismeasurement. The rapid growth of price for performance in the computer industry led the US Commerce Department to make an unprecedented change in the way that it calculated prices for hardware, which in turn can have a dramatic impact on the calculations that economists make as to the value of computer investment in the economy. Changing this number has, in turn, a dramatic impact on measures of total factor productivity—*especially* in industries that invested significantly in ICTs.

Usually, the Commerce Department measures price changes by looking at the price of a product last year (say, a packet of 2005 M&Ms), looking at the price of that same product the following year (a packet of 2006 M&Ms), and calculating the percentage difference. They make no accounting for the fact that in 2006, the M&Ms might be intrinsically worth more to the consumer because of a new color. Such an approach obviously has its problems in the area of computer products. It would be like comparing M&Ms to fine Belgian truffles to compare the average 1995 desktop PC to the average 2006 desktop PC, for example.

And so, the Commerce Department came up with a "hedonic" measure of computer prices. Put simply, it compared the price of a computer with a given set of characteristics (1mb hard drive, 16-inch monitor, and so on) to the price of a similar computer the year before. Such a system produced price indices that showed very rapid declines. Using $100 worth of 1996 computing hardware as a baseline, Robert Gordon found that the cost of such hardware declined from $61,640 in 1961 to $36 at the end of 1999.[25]

This rapid fall, in turn, can have a significant effect on investment figures. Economists should use "real," or inflation-adjusted, figures to look at growth in capital stocks and output over time. The US Commerce Department estimates that the hardware and software output of the computer and telecommunications industries equals about 8 percent of US GDP but has accounted for 25 percent of real growth since 1994. In nominal terms (not adjusted for inflation), however, it is responsible for just 10 percent of output growth. And from a 1996, inflation-adjusted perspective, a dollar invested in computers in 1999 bought three times as much "real" investment as a dollar invested in computers in 1996. Because of that, in 1996 "real" dollars, business investment in computers in 1999 was 60 percent of total investment, as compared to 35 percent in "nominal" terms. In short, these hedonic measures make ICT-producing industries look incredibly efficient and the ICT investment number look very large, as compared to traditional measures. (Their impact on the size of capital stocks is less dramatic because rapid improvement in quality for cost also signifies rapid capital depreciation.)

But hedonic measures are not free from controversy. They are not used in other sectors, despite some significant improvements in products over time (compare the Wright brothers' plane with a corporate jet). It is hard to know what impact hedonic improvement really has on an economy, especially if the hedonically improved goods are exported (a subject I return to in a later chapter). And such measures involve a somewhat arbitrary assumption about the "value" of computing power—that a computer with twice the megabytes should be valued twice as much. That point is arguable if there is a declining marginal return on computing power.

A 2000 *Economist* survey pointed out that a Ford Taurus car contains more computing power than the multi-million-dollar mainframe computers used in the Apollo space program.[26] And yet the Taurus still can't fly to the moon. Indeed, wouldn't most drivers rather have a Mustang (let alone just one moon rocket) than a Taurus? This is one example of diminishing returns on computing power. That power has become cheaper, so that rather than being used to calculate the relative velocities of a 3 times 10 to the 6th kilogram rocket, a 6 times 10 to the 24th kilogram planet, and a 7.35 times 10 to the 22nd kilogram moon and avoid the first crashing into either of the second two at 25,000 miles per hour, that same computing power is now used to tell you that you've left the trunk open.

A less prosaic example would involve the desktop word processor. In 1964, IBM brought out the magnetic tape/Selectric typewriter (MT/ST), which allowed entered material to be edited without being typed onto paper and provided electronic storage and replay, correction, and multiple reprinting. Over the next twenty years, world processing advanced at so rapid a pace that by 1986, there were more than sixty complete word processing systems on the market (including a very early version of Microsoft Word).

By the mid-1980s, the standard word processor already featured justification, alignment, insertion/overstrike, search and replace, copy/cutting, pagination, headers and footers, footnoting, table of contents and index generation, form letter merging, and automatic spell-check. Progress did not stop there, however. In 1989, Microsoft brought out its first edition of Word for Windows. It captured over 60 percent of the IBM-compatible word processing market within five years.[27] Microsoft Word, which now completely dominates the field, continues to be issued in newer versions with yet more features, requiring yet more powerful computers to operate it.

In 1995, Microsoft introduced what was already the eighth version of its word processing software (the grandfather of the version on which this book was written). The new version of Word—Word 95—had considerably higher requirements than its forebears: a 386 processor (with over 100,000 transistors on the chip), six megabytes of RAM, an 8-megabyte hard drive, and a 3.5-inch disk drive. To go along with this extra capacity, users were given a range of new features, including an "answer wizard" help system, auto-formatting, and designer templates.

It should be noted that a study of sixteen people using one version of this software found that of 642 commands available on the program, just twenty accounted for 90 percent of use. The average person used just fifty-seven commands in six months, and all the study subjects together used only 152 commands in eighteen months—or less than one-quarter of all those available.[28]

Nonetheless, that did not stop Microsoft issuing new versions with more commands. Within two years, the company had issued Word 97, which required a faster processor, more memory, a larger hard drive, and a CD-ROM. In return, users now had access to a letter wizard with layouts and styles, 3-D effects and animated text, and the animated smiling paper-clip office assistant.

Word 2000 came next. Again, system requirements increased. But users now had access to WYSIWYG (what you see is what you get) font menus and improved hyperlink functions. This was also the version that replaced the animated paper-clip office assistant with an animated Einstein office assistant.

Two years later, Word 2002 (or Word 11.0) required a Pentium 133, 32 megabytes of memory, and 375 megabytes on the hard drive. In return, users were rewarded with content-sensitive smart keys, faster organization of mass mailings (for which we can all be grateful), "a new auto-correct smart tag" that "lets you control how Word corrects text"—presumably allowing you to uncorrect all the mistakes made in auto-corrected text from the Word 2000 version—and an auto-save function that activates when the excessively baroque program crashes the computer.

Only a year later, Microsoft issued yet another version, Word 2003, again expanding system requirements in order to bring you Microsoft Clear Type®, which produces letter shapes that are easier to read. But still not content, the company has announced Word 2007, which will provide yet more features such as 3-D shapes (or, more precisely *better* 3-D shapes that you've had since Word 97) and improved corruption-recovery (when the auto-save from Word 2002 doesn't work in time). This latest version requires 256 megabytes of RAM and two gigabytes of space on your hard drive. That's a 42-fold increase in RAM and a 250-fold increase in hard drive requirements over Word 95.[29]

One wonders at the value added. Indeed, there are reasons to think that, by providing features that many users positively don't want, the net impact of new versions may actually be negative. James Fallows, a columnist in the *American Prospect*, suggests one such frustration:

> A homely example is the dancing paper clip, the "Office Assistant," in Microsoft Word. It pops up and moronically observes, "It looks like you're writing a letter!" when you type "Dear" and someone's name at the top of the page. If you complain to the company, as thousands do, its representatives act puzzled that you should consider this a problem. If you don't like it, why not turn it off?
>
> The immediate answer is that turning it off is not exactly an intuitive process. (1. Press the F1 key. 2. Press the "Options" button under the Office Assistant. 3. Make sure the box next to "using features more effectively" is

> unchecked—this controls the "Letter Wizard," though the menu says nothing about letters.) The larger point is that today's high-tech world is pulled in opposite directions about how much choice to ask of the user.[30]

In other words, the fact that my computer and the programs it runs have doubled in power and have twice the features in no sense makes it worth twice as much to me. Indeed, if it now runs excessively baroque programs, it may be worth less. Or, looking at it in the truly "hedonic" way, just because I would not pay more than $10 for a 1982 Sinclair Spectrum computer with 16k of memory, doesn't mean that my current 32 Mb machine with 2,000 times the memory is worth $20,000 to me.

Thus hedonic measures may overestimate the real value of output from the ICT-producing sector and also the real level of investment in ICT-using sectors. That, in turn, may affect the residual TFP calculation for both sectors, suggesting a slower growth in TFP in ICT-producing sectors and faster TFP growth in ICT-using sectors. Readjusting hedonic measures of price change covering ICT will somewhat change the overall output story (depressing it) but may more dramatically change relative TFP performance across sectors. It may be, then, that at least some of the underperformance of ICT-using sectors in TFP is due to an overestimate of their total investment in ICTs.

Baroque Networks

Despite this potential explanation for ICT investment's underperformance in terms of total factor returns, the cross-country and cross-state evidence presented earlier, showing no link between nominal levels of ICT investment and comparative growth rates, still suggests at the very least that ICT investment does not generate supra-normal returns. And the increasingly baroque redundancy of software, along with the attached unnecessary investment in hardware, may provide a partial explanation as well—or at least why we are not seeing a significant positive productivity impact through network externalities.

Backers say the Internet has a significant impact on TFP because the value of the network increases exponentially with each additional connection (one connection allows one e-mail route; two connections allow three routes; three connections allow six). That is the positive side of the story, but there are negatives to network externalities as well. If a partner company upgrades its software from Baroque 7.0 to Baroque 8.0 and staff at your company want to continue reading their files, you have to upgrade as well—whether or not that upgrade makes sense from an individual company perspective. Given the fact that people do not use the additional functionalities provided by the next release of Microsoft software—and have not for many releases already past—this network effect (rather than any real benefit to the individual consumer) might well be the driving force behind recent Microsoft Word sales. Within companies, it may be that, as network-based management systems are introduced, all staff have to be provided with network access—despite the fact that, from an individual productivity perspective, it only makes sense to provide half of the staff with computers. Within-company negative network externalities might reduce or remove TFP gains. Across-company negative network externalities might actually return TFP losses.[31]

Overall, then, it appears that there is some, although perhaps overinflated, evidence of strong TFP performance in production. Regarding use, it appears that the US evidence broadly favors the position outlined by Kevin Stiroh, at the New York Federal Reserve, in two papers published in 2002: in the first, he argues that industries that were intensive in their use of IT saw higher labor productivity growth, by as much as 2 percent. In the second, he concludes that "ICT capital appears correlated with the acceleration of average labor productivity (ALP) as predicted by a standard production model, but not with total factor productivity (TFP) growth," as would be implied by a new economy model.[32] (Again, labor productivity increases are the sum of contributions of greater capital per worker—investment—and technological progress—TFP.) More broadly, a recent OECD publication involving cross-country work carried out on its member countries suggests a result similar to that found in the United States: TFP returns on ICT manufacturing, labor productivity returns on ICT investment, and some at best mixed firm-level evidence of a TFP impact on ICT use.[33] The low impact of ICT investment on TFP may result from negative network externalities, mismeasurement of ICT capital stocks, or investment based on a misconception of ICTs as more revolutionary than they really are.

At the same time, it is important to remember the words of Richard Nelson from Chapter 2 regarding the weaknesses inherent in TFP calculations. Undoubtedly, technological advances in ICT have allowed for many billions of dollars of productive investment across the OECD, and this investment has increased output. Without advances in ICT, labor productivity growth would have been slower across Western economies in the 1990s. Nonetheless, ICT use did not create the type of returns and spillovers that would suggest we were in a new economy, or there was a market imperfection at work that saw ICT investment rates considerably below those that made economic sense.

Lagged Effects?

ICT proponents, faced with the evidence of little revolutionary impact to date, suggest that ICTs are "general purpose technologies" that can and will affect productivity across sectors, but historically, it has taken a considerable length of time for the impact from such technologies to be felt in economic statistics—as long as fifty years. Earlier examples of such general purpose technologies include the dynamo, where there was indeed about a fifty-year gap between introduction and ubiquity and impact, according to Paul David of Oxford University.[34]

From the point of view of developing countries, if we take today as year zero, this might suggest they should wait until the bleeding edge elements of technology adoption are over—perhaps another thirty years or so. But of course, this isn't year zero for ICT investment. In 1954, J. Lyons, a bakery firm in the UK, computerized wage and office calculation systems.[35] According to this start date, then, the computer's fifty years is up, and we should have seen the impact by now. Either way, pinning hopes for a significant productivity impact on one interpretation of the past impact of very different technologies seems a worryingly thin case.

Robert Gordon suggests a reason we may not be seeing a large impact from ICTs, compared to previous general purpose technologies. He agrees with proponents of the "lagged effects" thesis that there was a wave of increased growth based on TFP that peaked between 1928 and 1950. He also accepts that the peak period was built in part on the back of four great clusters of inventions—electric light and electric motors (and follow-on inventions such as air-conditioning and refrigeration); the internal combustion engine (and resulting highways and the supermarket); petroleum and petrochemicals (including plastics); and the complex of entertainment, communication, and information innovations developed before World War II (including the telephone, radio, movies, records, television, and mass-circulation newspapers). However, he argues that the ICT "revolution" is not as major a phenomenon as one of these clusters of inventions, let alone all four:

> Electric light . . . extended the length of the day. . . . the electric motor and the internal combustion engine created faster and more flexible movement, directly raising the productivity of factory workers, housewives, truck drivers. . . . refining, chemicals, plastics and pharmaceuticals all involve the physical rearrangement of molecules that change materials into more productive forms. The complex of electric and electronic entertainment and information industries arrived into a void in which nothing comparable had existed.[36]

Indeed, Gordon argues that the ICT cluster isn't even the most dramatic communications innovation of the last 200 years: "No current development in communications has achieved a change in communications speed comparable to the telegraph, which between 1840 and 1850 reduced elapsed time per word transmitted by a factor of three thousand (from ten days to five minutes for a one-page message between New York and Chicago), and the cost by a factor of 100."[37]

As we have seen, invented technologies such as ICTs are probably not the prime mover behind changes in TFP growth at the best of times, and if ICTs are not even a particularly dramatic form of general purpose technology, then it may not be surprising that we are struggling to find a uniquely positive economic impact.

The Economic Impact of E-Commerce

It is almost impossible to say anything about the Internet per se (as opposed to ICTs in general) from a macroeconomic standpoint—there is too little data. But e-mail has clearly saved billions in fax and phone costs as well as the time associated with multiple transmissions of the same document. And networking has greatly eased collaborative production of documents and software, for example. Internet-based e-transactions have also dramatically changed the nature of a number of commercial relationships.

Take the phenomenon that is eBay. Especially for niche products, it has been an incredibly powerful tool to bring together buyers and sellers. On any given day, according to company statistics, there are more than 12 million items listed on eBay across 18,000 categories. For example, you can buy a prepackaged church ministry website design, a personalized strawberry shortcake doorknob hanger, or a supervet calf snare.[38] In 2005, eBay members transacted $44.3 billion in gross merchandise sales (equivalent to around seven billion supervet calf snares).[39]

Internet-related technologies have, in some cases, had a dramatic impact on the efficiency with which tasks can get done, then. Some commentators, including Bill Gates, have suggested that the e-commerce phenomenon will lead to dramatic changes across the economy. Gates describes online shopping as "friction-free capitalism":

> The Internet will extend the electronic marketplace and become the ultimate go-between, the universal middleman. Servers distributed worldwide will accept bids, resolve offers into completed transactions, control authentication and security, and handle all the other functions of a marketplace, including the transfer of funds. We'll find ourselves in a new world of low-friction, low-overhead capitalism, in which market information will be plentiful and transaction costs low. It will be a shopper's heaven.[40]

That may somewhat overstate the case, which is perhaps lucky for Gates, who benefits from continuing market distortions.[41] The evidence for disintermediation (the removal of intermediaries between the initial supplier and final consumer of products—shops being one example) is far from clear, as John Seely Brown and Paul Duguid point out in their 2000 book, *The Social Life of Information*. Across the United States, the number of staff in administrative ("intermediary") positions is increasing—the growing number of nonproduction employees in manufacturing and the expansion of managerial employment as a percentage of the US workforce are both signs of this in action. And in the first six months of 1998, as pundits continued to predict the death of the conglomerate as falling transactions costs made smaller firms the norm, seven of the ten largest mergers in history took place.[42] Many other intermediaries are not disappearing but merely becoming monopolies—the browser wars between Netscape and Explorer were one example of that.

Overall, the Internet has not led to the blooming of a thousand viewpoints, or a thousand communities, but rather the development of remarkably few "supersites" that command mass attention. A recent survey of Web use reported that the top websites have a concentration of users even higher than their off-Web counterparts. For example, the top three news sites (MSNBC, CNN, and the *New York Times*) between them account for 72 percent of news site visits on the web. Weather.com accounts for 64 percent of all Web traffic toward weather sites. About half of Web users visited fewer than twenty different sites each month in 2001—a figure that is rising over time.[43]

Similarly, e-commerce has concentrated supply chains. For example, Dell, a company that has moved a good deal of its supply chain management online, has very few suppliers for a company its size, and thirty companies account for about 78 percent of total purchasing in 2001.[44] There was also a reduction in the number of General Electric's suppliers after it started its online bidding site for procurement.[45]

Perhaps because the Internet through e-commerce does not appear to have fostered some of the anticipated structural changes in the economy, it is as of yet hard to link the Internet itself with macroeconomic performance. The timing of the US productivity uptick was too early to be accounted for by the nascent tool of e-commerce.[46] In 1995, one estimate suggested that sales generated by the World Wide Web in that year were but $436 million.[47] Even in 1998, e-transactions were only worth $43 billion, or equivalent to 0.5 percent of US GDP.[48] And the slowing of growth in the nominal share of computers in the economy suggests some macro evidence of a declining marginal return on computer ownership actually coinciding with the expansion of the Internet. The nominal share of computer hardware in the economy rose rapidly *before* 1997. Growth since 1997 continued the unit-elastic response to the decline in computer prices prevalent prior to 1995: the share of nominal spending on computers was stable from 1997 to 2000.[49] Nominal spending then dropped by nearly 30 percent from October 2000 to October 2001 and has only slowly recovered after that.[50]

Furthermore, at the microlevel, savings from companies that have moved earlier and more aggressively to the Web have been far smaller than expected. IBM, for example, bought $27.7 billion worth of goods electronically in the first three quarters of 2000 and saved but $247 million (or a little under 1 percent) by doing so. On the sales side, 66 million customer service transactions were handled via IBM.com, but the headcount in call centers remained exactly the same.[51] More widely, only 40 percent of US companies that adopted online purchasing systems as part of a process of change management actually saved money—the figure for companies that introduced systems without a change management program was only 3 percent.[52]

Similar results have been found at the retail level. A recent review of studies on consumer prices on the Internet concludes the following:

> the Net has lowered most prices, but raised others. . . . Price dispersion is greater on the Net, and Net sellers change their prices more often. The role of intermediaries has not been reduced. The quality of service is idiosyncratic and highly variable. The Net is a new and distinctive avenue of commerce, but those who argue it has invalidated conventional economic principles are well off base.[53]

That is worrisome, considering many online sellers have been operating at such a significant loss (suggesting sustainable pricing might in fact be higher than in bricks-and-mortar stores).

Companies and individuals are not even trying to minimize their costs using the Internet. Only 3 percent of companies planned to use online reverse auctions in 2001 to guarantee the lowest product price for purchases, and between 55 percent and 89 percent of online shoppers purchase products at the first website they visit, for example.[54]

Banking provides an interesting case study here. Cap, Ernst, and Young estimates that rather than the predicted 25 percent savings in UK bank costs thanks to Internet banking, real savings in 1999 were on the order of 0.1 percent. And in markets like banking, where customer loyalty and high fixed costs of entry ($145 million for one British Internet banking venture) are important, new Internet-based companies have struggled. A range of companies from around the world (for example, Wingspan in the United States, First-e in Ireland, Marbles and Egg in the UK, and Sanwa Bank in Japan) have cut back, delayed, cancelled, or restructured their online banking ventures.[55] Not only are these problems likely to reduce the share taken by new Internet-based companies, but also the replication of fixed costs may reduce the impact of lower marginal costs on economic efficiency and growth rates.

More sober evaluations of the real potential gains from e-commerce might also account for the slowdown in e-commerce growth—between 1998 and the end of 2000, the FEI/Duke Corporate Outlook Survey suggests that the percentage of sales made online by surveyed companies remained unchanged at 5 percent.[56] More recently, the Census Bureau estimated total e-commerce transactions at $1.679 trillion in 2003. Two things are worth noting about this statistic. First, it was nearly all in the business-to-business (B2B) sector. E-commerce accounts for a healthy 19 percent of B2B shipments, sales, and (other) revenues but only 1.2 percent of total business-to-consumer (B2C) shipments, sales, and revenues. Compare that to the share of mail-order catalogs in retail sales, which remains over 10 percent.[57] Second, the majority of B2B traffic was not over the Web but over electronic data interchange (EDI) systems that far predate it.[58]

EDI systems use proprietary software to connect purchasers' and suppliers' computers to automate transaction and processing exchange and have been around since at least the early 1970s. Although Internet-based systems are estimated to have operating costs of about 1 percent of EDI systems, the cost of conversion to Internet-based systems is itself high enough to have discouraged the bulk of EDI users from so far switching to the Internet.[59] Thus the Internet offers only a marginal improvement on EDI, and the extent that B2B will spread beyond companies already connected by EDI will be limited.

At the macrolevel, returns might be even lower than company-level returns because of two negative externalities linked to e-commerce sites. First, the Internet allows businesses to force other companies and individuals to perform part of the service previously provided by the newly networked business. For example, Expedia.com encourages the customer to search for cheaper flights through the air ticket database rather than its travel agents. That saves Expedia money but does not necessarily increase economic efficiency. Second, and perhaps of greater concern to LDC companies with less access to the new technologies, some Internet investment (by, for example, Borders.com) involves defending market share (against Amazon.com)—so that social returns on this investment are lower than private returns.

The Economic Impact of Other Internet Applications

Of course, e-commerce is only one business use of the Internet. Undoubtedly, the major applications have been e-mailing and general World Wide Web use from the desktop. The average US employee spends a considerable amount of time online. In 2001, more than half of the US workforce over twenty-five used a computer at work, and 42 percent of them had access to the Internet. The average work user was online 2.6 hours a day. Some of that time was spent e-mailing distant colleagues rather than three more expensive options: calling (which leaves no written record), faxing (no electronic record), or physical mailing (which takes a good deal longer). The Internet also allows real-time distant collaboration in a way that no previous technology has done. As a result, it is perhaps unsurprising that, if you ask people about the effect the Internet has had on their work, more than 50 percent say that it has made them more productive, compared to 5 percent who say that it has made them less productive.[60]

Nonetheless, here, too, are reasons to think that putting the Internet on the majority of desktops in the US workplace is unlikely to have massively increased productivity. One major reason is that, for all that the Internet eases interaction between employees within and between firms, it is also one of the greatest time-wasting devices in history.

Ninety percent of US workers admit to surfing recreational sites during office hours (even admitting that nearly one-third of their time on the Net is spent on recreational surfing) and 84 percent of workers say they send personal e-mail from work. More than half of US workers admit to cyber-shopping on company time, and one in five online game players logs on from work. Twenty-five percent of employees with access feel that they are addicted to Internet usage at work.[61]

The data suggest, unsurprisingly, that the average employee is behaving even worse than he or she will admit. Nielsen's NetRatings (netratings.com) told us some interesting things about what Americans did with the Internet at work and at home. Perhaps the most interesting thing is that they did pretty much the *same* thing with the Internet at work and at home. Nielsen measured which parent companies were receiving hits on their websites and ranked them by unique audience. At home, Microsoft came first, with 59 million unique audience visits per week. At work, Microsoft came first again, with 38 million unique audience visits.

But it is not only the number one spot that was the same at home and work. Nine out of the top ten sites and nineteen out of the top twenty-five sites got similar numbers of hits, whether visitors were at work or at home. That's not just search sites or AOL. In terms of unique visits, eBay came fifth at home and sixth at work. Amazon was eighth at home and fifth at work. Walt Disney came in fourteenth at home and thirteenth at work. There are some differences—more news sites at work, more time spent looking for old classmates at home—but the similarities outweighed them.

There are two ways of looking at these results. One is to argue that similar Internet usage at home and at work suggests that people are being more productive at home—and that's probably true. But the more significant impact is fairly evident—people use the Internet at work as a (perhaps not terribly productive) source of entertainment. And given that in June 2002, the average home user spent less than half of the hours on the Web that the average work user did, they were spending much more time using the Web as a source of entertainment while they were being paid than they were when they were on their own time.[62]

Further, if it is true that they surf the same sites on the Net at work as they do at home, we know the *really* big thing many of them are doing. The top uncensored search terms give away a lot—"sex" comes first, "porn" second, and "Paris Hilton" third.[63] Sex is more popular than games, music, travel, jokes, cars, jobs, health, and weather combined. Danni Ashe is a paragon example in this area. Many readers of this book will not know who Ashe is, although the fact that her image has been downloaded more than a billion times suggests that some might. Suffice it to say that little bandwidth is used to download the pixels that make up the image of her clothing. Ashe, according to her press release, is twice as popular on the Web as Martha Stewart and three times more popular than Oprah Winfrey or Britney Spears.[64]

Twenty-five percent of surfers visit one of about 30,000 to 40,000 adult sites (there were 27.5 million visitors in January 2002 alone), but again, that underestimates the importance of this segment of the market. Because these sites are "content rich" (read: lots of pictures), they consume a huge proportion of network bandwidth—estimates suggest between 40 and 80 percent of total data traffic is porn.

Further, sex sites led the way in profitability on the Web—in 1999, adult sites took in 69 percent of all premium-content revenues on the Internet; by 2005, online porn's estimated $12 billion in annual revenues in the United States equaled the combined income of three major US networks (ABC, CBS, and NBC).[65] America Online's early success against competitor Prodigy has been put down to AOL allowing sex-related chat rooms where Prodigy banned them—half the chat rooms on AOL were sex-related in 1996, and they brought in $80 million a year in revenues.[66] And they have weathered the dot-com collapse better, as well—one hosting company reported that, over six months in the middle of 2001, 20 percent of mainstream customers went bankrupt, compared to 2 percent of porn sites.[67] Probably the only source of e-mail traffic larger than sex is spam—although, of course, a good deal of spam is about sex.

Again, much, if not most, of this surfing occurs at work. According to one recent survey, 37 percent of employees admit to shopping at work, but only 2 percent admit to viewing porn at work. Interestingly, when sites visited by employees are tracked, it appears that close to 70 percent of employees visit retail sites, with 60 percent of consumer online dollars spent at work,[68] and 37 percent check out X-rated sites. No doubt they arrived there by accident.

This amounts to a lot of futzing. If you add up the total hours that people spend at eBay, Amazon, and Walt Disney at work each week in the United States, it comes to 6.7 million hours. To put that in context, the working-age population of the United States is about 174 million, so every week, assuming they're all working 40 hours, the US spends about 7 billion hours at work.[69] About one-tenth of 1 percent of that time is spent on eBay, Amazon, and Walt Disney websites—not an insignificant amount for only three websites.

Taking the broader picture, about 47.5 million people have access to the Internet at work in the United States. On average, they spend seven hours and fifteen minutes on the World Wide Web each week. We have seen there are strong reasons to think a lot of that is wasted time from the point of view of employers. Even employees admit that one-third of the time they spent on the

Web is recreational. Let us estimate that the real figure is one-half.[70] That is 172 million hours a week, or 2.5 percent of total US working hours. Worst, it is the working hours of the better educated workforce (74 percent of people with an advanced degree use the Internet at work, compared to 8 percent of those with less than high school education) and the better paid (over half earn more than $50,000, and over 35 percent more than $75,000).[71]

If people are really spending a good deal of their time on the Web at work, and a good deal of that time has nothing to do with producing more widgets for the company, that might suggest a lower—or even negative—impact of Internet investments on median employer productivity. Given that World Wide Web usage patterns are the same at work and at home, one has to believe in very high returns on the time that individuals are actually using the Web for work purposes to think that the Web in particular (rather than the Internet in general) is a worthwhile addition to the productivity-enhancing tools of the average worker.

Not only that, a considerable percentage of company infrastructure expenses will be accounted for by the demands of this futzing. Some significant proportion of workplace data traffic is pornography. And even when employees are only checking e-mail, a good deal is personal, and, by now, over 40 percent is probably spam. According to several studies, spam cost US organizations over $10 billion in lost productivity in 2003.[72] One wonders about the return on a company's investment in a broadband connection if that is the case.

It should be noted that some large proportion of time spent at websites and e-mail unrelated to work will involve "substitute futzing"—people are going on the Web rather than playing Minesweeper on the computer, reading *Playboy* in hard copy, or staring out the window. But if the Internet is a more enjoyable futz than staring out the window, its probably safe to assume that total futz time has gone up, perhaps significantly, as a result of the Internet. Further, companies don't have to pay for employees' copies of *Playboy* or the view out of the window. They do have to pay for the infrastructure and services that bring Busty la Rue to the desktop screen. The silver lining? As noted by Dan Hess, the vice president of comScore Networks, "What many employers undoubtedly fear as an enormous productivity gap offers a substantial opportunity for marketers and advertisers in the targeting of online ads and other promotions."[73]

Implications for Developing Countries

What did we learn from the bursting of the Internet bubble in the United States? That the new economy wasn't all that different from the old, that—a few radical departures such as eBay aside—the new economy was the old economy with a new niche distribution and information channel, that scale economies and first mover advantages had a role in some parts of the economy (as before) but didn't in much of the rest, and that perfect markets and perfect information could not be formed at the click of a mouse.

That is, of course, a disappointment for those seeing the Internet as a major source of growth for developing country economies. If the benefits of TFP growth in particular accrue largely to production, it is a problem for a group of countries with comparatively small ICT industries (as we shall see). If there is limited evidence of a past spillover impact of investment in computers and the Internet on economic growth in the United States, and the case for a dramatic increase in productivity in the future is at least mixed, that suggests any benefits from the Internet in LDCs are likely to be greatly delayed and comparatively small. Again, it is worth noting that ICT investments did garner returns and create new avenues for profitable activities. They have done (and will continue to do) the same in developing countries. But changes in developing countries are likely to be slower and less dramatic than in the United States.

Furthermore, the changes that took place in OECD economies suggest a number of concerns for LDCs. We have seen that the Internet appears to concentrate opportunities among fewer suppliers. Microsoft's small business edition of its Office 2000™ Suite is an example of a technology that might further the process of supply chain concentration. The software has a built-in link to Office Depot stores worldwide as part of the office management system. Customizing the software to order from locally owned companies is far more challenging than accepting this default option.[74] If that kind of practice becomes more widespread on a global level, those companies least equipped to move online are likely to lose market share to those better placed. It is likely that the least-equipped companies will be concentrated in the developing world, where the physical, human, and institutional capital to successfully exploit the Internet is lower.

Indeed, a recent report by the OECD, *The Economic Impact of ICT,* concluded that successful adoption of ICTs by companies was determined in part by the cost of the technologies themselves, but also the availability and expansion of know-how and skilled operators, the scope for organizational change, and the capacity of firms to innovate including the status of the regulatory environment.[75] However, we shall see that although LDCs have a comparative abundance of "low-technology" ICT capital stock, both physical (television and radios) and human (literacy), they have much less "high-technology" capital (computers and tertiary education). Many of the high-technology institutional structures to maximize the benefits of ICT investment are also missing.

A particularly worrying trend involving ICTs is the increased need for employees with advanced education. The high human capital requirements of computer use in particular appear to have supported a trend toward an increasing divergence in wages between the university-educated and those who did not receive tertiary education. Particularly since the 1970s, the United States has seen a rapid increase in the demand for skilled workers, skill upgrading within firms, and a fall in the demand for unskilled labor.[76] The real wages of young men with twelve or fewer years of education *fell* by 26 percent between 1979 and 1993 and have not recovered since.[77] Similar patterns have already been found in LDCs. During the 1990s, the wage gap between unskilled and skilled increased 30 percent in Peru, 20 percent in Colombia, and 25 percent in Mexico.[78]

Skill upgrading in the United States has been apparent across sectors, but it has been even faster in computer-intensive industries than in others, and computer investment by industries is followed by increased employment of college graduates.[79] That trend should concern developing countries because, as we will see, it is precisely in the field of tertiary education that the gap in human capital stocks is at its largest between developed and developing countries.

It is worth noting again that skill upgrading and increased inequality are not necessarily associated with increased total factor productivity. These results have also been found in developing country contexts: a 1996 study of manufacturing firms in northern India found that IT investments were correlated with more highly skilled workforces but not with wages or productivity.[80] Given the lack of a clear link with productivity, it is perhaps not surprising that a number of studies suggest that pay differences between computer users and other, similarly educated employees reflect other factors than the value of computer skills. As an example, a recent German study found that the income gap between people who used a pencil at work and those who did not was apparently of the same magnitude as the income gap between computer users and nonusers.[81] Nonetheless, the increasing number of computers in firms has fostered demand for highly educated employees, which has driven up the cost of hiring college-educated staff.

Hardware and software upgrading presents an additional set of problems. Baroque networks may in particular be a potential force for divergence of incomes across countries. If developing world companies and individuals have even less need for baroque technologies than developed world companies, they will be the biggest losers from related negative externalities. I return to this idea in later chapters.

A further concern is that some innovations such as just-in-time management, assisted by ICTs, actually increase the cost of physical distance in manufacturing. That may be why, since 1985, the average distance between supplier plants and assembly plants in the US auto industry has actually fallen. Suppliers to Toyota, a leader in just-in-time techniques, are an average of just 59 miles away from the assembly plant, compared to those for General Motors (with inventories four times as big as a percentage of sales), where the average distance is 427 miles.[82] Quite the opposite of the suggested "death of distance," ICTs, which are playing a part in the just-in-time revolution, may be promoting an increased tyranny of distance, at least in some industries.

The remaining, or reinforced, tyranny of distance, combined with the skills and institutions required to successfully use the Internet, may be one factor explaining results from a number of studies based on US and European data that suggest that backward companies and regions are benefiting less from the Internet than already better-off regions or productive companies. Those firms benefiting most from the introduction of IT were already thriving in the "old economy."[83]

Proponents still make much of the potential for LDCs to trade more and different products due to the death of distance in service industries, at least. Tony Venables of the London School of Economics suggests that these opportunities are much overrated.[84] Many "knowledge goods" remain embodied in human or physical capital that is still expensive (and difficult) to transport across international boundaries. Venables also notes that as goods become weightless, they also tend to be subject to dramatic productivity increases and price reductions. Taking the example of airline ticketing, he notes that the major impact of ICT has been to replace labor with computer equipment—and only secondarily to allow remaining workers to be employed in remote locations.

Looking at the type of "low-tech" service jobs that *are* now exportable, it is doubtful that there are enough to make a significant difference to LDCs as a whole. Data entry (the low-skills end of the information processing sector) is a $800 million industry in the United States. Imagine that the United States has only its share in global GDP of the data entry market (about 27 percent of the world total)—so that, worldwide, the industry is worth approximately $3 billion. That is equal to a little less than the yearly exports of Estonia.[85] More generously, the International Labour Organization (ILO) reports that perhaps 5 percent of all service sector jobs in industrial countries are "contestable" by LDCs[86]—nonetheless, that totals a mere 12 million jobs, or 0.24 percent of the population of the developing world.[87] In 2003, the actual number of service jobs outsourced overseas was probably closer to 1.5 million.[88]

Worse, even in services, the big opportunities appear to be going to the already wealthy. Production of high-value-added knowledge goods, in particular, has remained concentrated in wealthier regions, despite falling communications costs.[89] There is global evidence that dot-com firms are also highly concentrated in regions that were wealthy prior to their creation.[90] If anything showed the death of the "death of distance" in software, it was Microsoft relocating part of its research operation to Silicon Valley from Washington state in 1999 because it was fearful that physical distance from the center of Internet creativity might dampen growth prospects.[91] More broadly, the biggest recipient of global outsourcing in 2002 was the United States, providing more than three times the business services exports of India and China combined.[92]

Finally, the high risk of ICT investments should worry developing countries. The average investment in developing countries already carries a far higher risk than the average investment in the OECD. The complexity of ICT investments only adds to this problem. We have seen evidence of high failure rates in OECD countries. In LDCs, with less suitable environments for implementation, Richard Heeks estimates that as many as 80 percent of IT projects fail to meet all their objectives.[93] If this figure is even close to accurate, it suggests the caution with which such investments should be made—and the very high returns that should be expected. In particular, projects appear to be at risk when not carried out as a larger process of change in institutional structures. ICTs are largely a supporting tool for greater efficiency—rarely does wiring old processes lead to significant savings, as we have seen. It is clear that the importance of the broader environment in which ICT investments are made is central.

On the plus side, there is likely to be a comparatively limited spread of Internet-enabled computers within connected companies in LDCs. That will, hopefully, limit access to those employees who will see the highest returns from use while limiting the hours lost to Internet "futzing." Even those with Internet access are likely to be less prone to waste time online, because World Wide Web access is often frustratingly slow in developing country companies.

Conclusion

Overall, US evidence suggests that ICTs have been a powerful economic force for growth. Investments in ICTs were a significant cause of labor productivity growth in the 1990s. The ICT industry was a significant source of total factor productivity. However, there is little or no evidence that ICT investment generated especially strong returns, for reasons including the negative externalities of baroque technology advance. And the nature of economic changes in which ICTs did play a role—in production management, for example—suggests that developing countries might see downsides from the ICT revolution, much as a company that is forced to upgrade its software because its clients have upgraded—the company spends money for little financial benefit. Developing countries face significant challenges in terms of the human and physical capital requirements for widespread Internet use. Beyond the nature of the Internet as an embodied technology, it also appears to carry significant institutional demands that make e-development a complex and risk-prone venture, even in rich-country settings. Evidence from wealthy countries suggests the result that we would expect, given the theory presented in Chapter 2—ICTs and the Internet to date have been comparatively minor sources of growth, but to the extent they have had an impact, they have been forces for divergence in incomes and concentration of economic opportunity. Institutional settings and factor endowments do appear to matter for the exploitation of the technology of the Internet, and it may be that rich countries and regions are advantaged in this regard. We will look at these issues in relation to LDCs in Chapter 3.

Notes

1. Cited in Berman 1988: 21.
2. Cassidy 2002.
3. US Department of Commerce 1999.
4. White House 2000.
5. Oppenheimer 1997.
6. Robert Solow, "We'd Better Watch Out," *New York Times Book Review*, July 12, 1987. Solow took it back in 2000 (see L. Uchitelle, "Economic View: Productivity Finally Shows the Impact of Computers," *New York Times*, March 12, 2000, section 3, p. 4).
7. A recent "survey of surveys" concluded that in the 1980s and the early 1990s, the consistent finding was that there was a broadly negative correlation between IT investment and economy-wide productivity in the United States—the "Solow Paradox" mentioned above. A few studies in the late 1990s began to reverse this conclusion, however, as investment in computers expanded (Pohjola 1998).
8. Gordon 2003.
9. Although it should be noted at the same time that companies were investing less than previously in everything else—"capital shallowing" in non-ICT investments accounted for –0.1 percent of labor productivity growth. See Perkins and Perkins 1999b.
10. Gordon 2003.
11. McKinsey and Company 2001.
12. Baily 2002.
13. Lehr and Lichtenberg 1999.
14. David 2000. Greenan, Mairess and Topiol-Bensaid 2001 also find no evidence that increasing computerization and R&D over time has a significant impact on firm performance.
15. Schreyer 2000. Harvard economist Martin Feldstein (2003), points out that there are a number of other reasons for variations in total factor productivity growth between the United States and Europe in the late 1990s, including (he argues) the greater use of stock-based compensation in the United States than in Europe.
16. Bank of England 1999.
17. Gordon 2002.
18. "Solving the Paradox," *Economist*, September 23, 2000. See also Stiroh 1998 for a similar result.
19. US Department of Commerce 1999.
20. "The New 'New Economy,'" *Economist*, September 11, 2003.
21. Gordon 2002.
22. Brown and Duguid 2000.
23. See www.standishgroup.com/sample_research/chaos_1994_1.php.
24. Triplett 1999.

25. Gordon 2000a. Hedonic price indexes have a number of problems. For example, constructing hedonic indexes based on the set of computers available for sale produces figures of a 30 percent decline in price per year from 1985 to 1991. Constructing the index using data on systems actually purchased produces a 20 percent per year rate of decline (see Brown 2000).

26. "The New Economy," *Economist,* September 23, 2000.

27. Information in this paragraph came from Kunde 1996.

28. Linton, Charron, and Joy 1998.

29. The post-1995 information on MS Word all came from the Microsoft website: www.microsoft.com/ms.htm.

30. Fallows 2000.

31. Morrison and Siegel's (1997) paper looking at manufacturing in the United States found that computer investment in one subindustry was actually associated with greater productivity in other industries in the same sector. It is worth noting, however, that this paper discusses labor rather than total factor productivity.

32. Stiroh 2002a, 2002b.

33. OECD 2004a. See also Daveri and Silva 2004, who argue that in Finland, TFP growth gains outside Nokia and a few other IT-related service industries have been temporary or nonexistent.

34. David 1990.

35. "Is this the first step in an accounting revolution or merely an interesting and expensive experiment?" asked the *Economist* in an article on the Electronic Abacus, March 13, 1954.

36. Gordon 2000a.

37. Gordon 2000b.

38. All available on www.ebay.com on April 28, 2003.

39. http://pages.ebay.com/community/aboutebay/overview/index.html, accessed April 28, 2003 and May 10, 2006.

40. Excerpt from http://www.roadahead.com/, the website of *The Road Ahead*, Bill Gates's first book.

41. "The New Geography of the IT Industry," *Economist*, July 21, 2003. Between them, Office and Windows generate $20 billion in revenues with a margin of over 80 percent, suggesting that network effects and intellectual property protection between them may be crimping the effects of competition somewhat.

42. Meanwhile, as the paperless office took hold, US paper consumption rose from 87 to 99 million tons a year in the 1990s. Paper consumption per head in offices has doubled since 1975. The average office worker consumes more than his or her weight in paper each year at work. See Brown and Duguid 2000.

43. Reported in the *New York Times*, August 29, 2001.

44. "Inside the Machine," *Economist*, November 11, 2000.

45. World Bank 2000b.

46. Contra Nezu 2000.

47. From Nua, at www.nua.ie.

48. "How Real Is the New Economy?" *Economist*, July 24, 1999.

49. Gordon 2000a.

50. DeLong 2002.

51. Roberti 2001.

52. Yusuf 2004.

53. Koch and Cebula 2002.

54. Baker et al. 2001.

55. "The Hollow Promise of Internet Banking," *Economist*, November 9, 2000.

56. www.duke.edu/~jgraham.

57. Almasy and Wise 2000.

58. US Census Bureau 2005. Total e-commerce sales for 2005 were estimated at $86.3 billion, an increase of 24.6 percent from 2004. E-commerce sales in 2005 accounted for 2.3 percent of total sales. "Quarterly Retail E-Commerce Sales, 4th Quarter 2005," February 17, 2006. www.census.gov. 77 percent of those people who do shop online make fewer than ten purchases a year. Only 16 percent of those people think that Internet prices are lower than those in traditional retail stores.

59. World Bank 2000b.

60. UCLA 2001.

61. Naughton 1999; eMarketer 2003. That is 10 percent of the total US workforce that need an ICT equivalent of Methadone (Minitel, perhaps?).

62. www.NetRatings.com

63. www.wordtracker.com. Lycos provides what looks to be a cleaned-up version of its top fifty search terms every week, in which sex and porn don't even make an appearance. But the terms that do still don't make comfortable reading for those linking the Internet with productivity: Paris Hilton, Pamela Anderson, Poker, NFL, Pokemon, Apple, Spyware, LimeWire. None of the top fifty search terms looked any more work-related than file-swapping programs, and it is a fair bet that they appear not because people wanted to swap Powerpoint presentations over the Web (see http://50.lycos.com; these statistics were for the week ending June 3, 2006).

64. www.billiondownloadwoman.com. The top porn site in the United States, as of 2001, was called Karasxxx.com and had 6.9 million users each month.

65. A report of the Third Way, which represents the progressive bloc of the Democratic Party in the United States showed this finding. Quoted in Philip M. Lustre, "Pornography the Bane of Broadband Revolution," *Manila Standard,* January 12, 2006.

66. Quoted in Cassidy 2002.

67. See Perdue 2001; *USA Today*, February 26, 2002; Nielsen NetRatings (NetRatings.com); Alexa Research (www.alexa.com); and M. Brunker, "Sultans of Smut," on msnbc.com.

68. eMarketer 2003.

69. http://www.ameristat.org

70. Why one-half? UCLA's (2001) survey of Internet use suggests that Internet users spend about 22 to 23 percent of their time online on e-mail, 3.4 to 6 percent on news, and 2.7 to 8.2 percent on "professional work." Even assuming *all* the e-mail and news use is work-related, that adds up to 37 percent of time spent online being work-related. One would hope (although there's not much evidence) that people spend more time at work on work-related activities than they do at home. Even given that, 50 percent seems a generous estimate of time spent online in work-related activities. As further evidence, recent monitoring of the Internal Revenue Service (IRS) workforce's Internet use found that activities such as personal e-mail, online chats, shopping, and checking personal finances and stocks accounted for 51 percent of employees' time spent online (see Davis 2001).

71. eMarketer 2003.

72. Between the start of 2002 and the start of 2003, the percentage of spam as a proportion of e-mail jumped from around 17 percent to around 40 percent. "Stopping Spam," *Economist*, April 26, 2003. According to the Web research site, Top Ten Reviews, spam is likely to increase by 63 percent by 2007. "LinuxForce (SM) Releases

Version 3.0 of Acclaimed Anti-Spam, Anti-Virus Service," *Business Wire,* April 4, 2006. http://www.businesswire.com.

73. eMarketer 2003.

74. Gomez 2000.

75. OECD 2004. See also Bresnahan, Brynjolfsson, and Hitt 2002; van Ark and Piatowski 2004; and the review in Indjikian and Siegel 2005.

76. Autor, Katz, and Kreuger 1998.

77. Berman, Bound, and Machin 1998.

78. Oxfam 2000.

79. Autor, Katz, and Kreuger 1998.

80. Lal 1996.

81. Bedi 1999.

82. Venables 2001.

83. World Bank 2000b; Doms, Dunne, and Troske 1997; Greenan, Mairess, and Topiol-Bensaid 2001.

84. Venables 2001.

85. Schware and Hume 1996; World Bank 2000c.

86. ILO 2001.

87. Calculated from World Bank 2001a.

88. "Outsourcing: Getting the Measure of It," *Economist*, June 30, 2005.

89. Cornford 2001.

90. Gillespie, Richardson, and Cornford 2001.

91. Brown and Duguid 2000.

92. Amiti and Wei 2004.

93. Heeks 1999a.

4

ICT in the Developing World

BASED ON WHAT WE have learned about the impact of technologies on growth in theory, as well as the evidence of the impact of ICTs and the Internet on economic performance in rich countries, we can now turn to developing countries. In this chapter I examine the potential benefit to developing countries of the significant TFP gains that can be garnered in the production of ICTs. I also look at the evidence regarding rollout and use of ICTs in general and the Internet in particular in developing countries and examine the reasons behind limited advanced use in particular. As expected, I find that the broader environment for exploiting e-commerce is not there, which points to the Internet as a potential force for divergence in incomes. I conclude the chapter with some of the more positive estimates made of the impact of the Internet in developing countries and note that, even were such gains to be accomplished, they would still represent a comparatively small step on the path to high-income status.

Production of ICTs in Developing Countries: Wealth for Some?

We have seen that if there is a significant TFP effect related to the ICT revolution in the industrial world, it is on the production side that the effect is noticeable in the statistics. At first glance, that suggests a significant opportunity for developing countries. And, indeed, a number of developing countries around the world *have* seen significant investment, employment, and export growth thanks to ICT production. Between 1994 and 2000, electronics exports from the state of Guadalajara in Mexico increased to over US$5 billion per year, and the industry grew to employ over 60,000 workers.[1] India exported

$12 billion in software and services in 2005, with the software and outsourcing industries employing perhaps 800,000 people.[2] The country's IT industry represents nearly 3 percent of GDP.[3] Twenty-one percent of Malaysia's GDP is accounted for by production of ICTs, and 28 percent of East Asia's manufactured exports are ICTs.[4]

We saw in an earlier chapter that the estimates of significant TFP gains in the sector were the results of greater power-for-cost performance of the sector's products (although we saw that the "real hedonic" benefits of these improvements may be overreported by such statistics). And at least one study has suggested some considerable TFP impact in some of the developing countries with a significant ICT-producing sector, concentrated in East Asia. Using the same techniques we discussed earlier for measuring the impact of that sector in the United States, the authors suggest annual TFP increases due to ICT production over the 1995–2000 period of 7.04 percent in Singapore, 3.47 percent in Malaysia, and 1.9 percent in Taiwan and Thailand—suggesting impacts far larger than felt in the United States itself.[5]

Overall, however, LDCs are largely importing goods in the IT sectors, not inventing or even producing them. The ITU estimates that exports of telecommunications equipment are worth only 8 percent of imports of such equipment in low-income countries and 40 percent of imports in middle-income countries.[6] Low-income countries are responsible for only 0.3 percent of the world's high-technology exports.

Given how small IT sectors are in most developing countries, it is unlikely that they will have a large impact on growth. One recent study estimates the impact of TFP growth in the ICT-producing sector of India added only 0.05 percentage points of growth to the Indian economy from 1995 to 1999: this in the country trumpeted as a model for ICT-production impact in the poorer parts of the developing world.[7] Furthermore, the TFP gains from IT companies that have occurred in developing countries are unlikely to return the kind of impact on real economic wealth that TFP gains in the United States might bring.

Doubtless the East Asian tigers are the countries in the developing world where, in percentage terms at least, the ICT revolution has had the largest impact on domestic production and employment. Having said that, it is also likely that the TFP numbers suggested above overestimate the impact on people in Singapore, Taiwan, or Malaysia—even those who work in or own ICT factories and even more than hedonic measures may overestimate the impact in the United States.

Who benefits from the hedonic gains? In some cases, such as Microsoft, we've seen that the producer can hold on to TFP gains to make profits and pay large salaries. But most of the time, we would expect (in competitive markets) the benefits to go to the consumer. We can see that that is true because the price of computers hasn't risen as they have become more powerful. Instead, the consumer gets more bang (more 3-D playability in their children's shoot-

'em-up, faster processing of their tax program) for the same buck. If the consumer gets this benefit, then the producer does not. And if the consumer sits in the United States, while the producer sits in Taiwan, that greater power—the hedonic gain and the TFP increase—benefits the US economy, not the Taiwanese economy. Putting it another way, the productivity increases in the ICT sector have led to rapidly declining terms of trade for (quality-adjusted) output of ICTs. East Asia is producing more computer power at the same cost, but the computer power sells for less—so the net impact on East Asia is zero.

Developing country producers in competitive, commoditized ICT sectors do not see very significant returns from increasing "hedonic productivity": one estimate suggests US companies produced 56 percent of the revenues yet garnered 96 percent of the profits from the global IT industry in the late 1990s.[8] The United States dominates patent ownership, where a publicly enforced monopoly allows technology creators to garner the returns from technology advance. The rest of the world, much more involved in licensed production, cannot protect monopoly profits and so is forced to pass on the lower cost of ICT capacity to (largely rich-country) consumers.

Developing countries are now patent consumers (licensees in this case), and it is unlikely they will become patent producers (licensors) any time soon. LDCs spend far less on developing new technologies than industrialized countries do. Expenditure on R&D in low-income countries combined totaled approximately US$5 billion in 1999, compared to the figure for the United States alone of US$234 billion. Not surprisingly, those figures translate into OECD dominance of world patent applications. In 1998, 1,114,408 patent applications were filed in low-income countries. Fewer than 10,000 of these applications—or under 1 percent—were filed by residents. In turn, royalty and license fee payments by low-income countries were nine times royalty receipts, whereas in the United States, royalty receipts were 2.7 times payments.[9] Perhaps the dissipation of productivity impacts from IT production explains why East Asia, the developing region with the largest IT industry, sees no correlation between high-tech exports (as a proportion of total exports) and total productivity growth measures.[10]

Looking at the case of India in particular, we see a broader lack of competitiveness means that the country will face difficulty competing further up the "value chain" of ICT production toward patentable outputs. Studying a range of skilled labor costs in the IT industry shows India's average costs (in categories such as systems analyst and support programmer) are about one-half of Ireland's labor cost per annum. Compare that to revenues per employee of about one-quarter Ireland's level.[11] For Indian IT firms to remain profitable, either capital costs must remain low, with a continued focus on "onsite services," which account for over one-half of the country's IT exports,[12] or the percentage of unskilled to skilled labor in the firm must remain very high. Neither solution is attractive if the aim is to move up the value chain.

Furthermore, India remains one of the developing countries best placed to benefit from ICT production. Richard Heeks and Brian Nicholson's survey of software export success factors in India, Israel, and Ireland concludes that the three countries shared advantages that, sadly, are not widespread in the developing world. These advantages included high human capital stocks available at low costs, significant diaspora populations, access to technology (in part built on significant defense industries), widespread use of English, and strong ICT networks. (The countries also passed strong intellectual property rights [IPR] legislation and used significant tax breaks and incentives to encourage R&D, marketing, and industry growth—suggesting that the economic rate of return of the industries created may well be below the private rate of return.)[13]

Overall, then, a few developing countries have and a few more may develop significant ICT-producing sectors. But even those lucky few may see only a "normal" growth benefit from their efforts—similar to production of steel, cars, or widgets. There is little reason to believe that the ICT revolution will bring supra-normal returns to producers in developing countries; nor will the majority of developing countries see particularly rapid growth of ICT industries.

The Use of ICTs in Developing Countries: The Shrinking Digital Divide

One common reason for fearing that the Internet will have differential impacts in rich and poor countries—the so-called digital divide—may be overplayed. Furthermore, the recent rollout of mobile services may well have had a significant economic effect. Nonetheless, we shall see that differences in use (as opposed to access) suggest that the Internet may well have a smaller impact in developing countries than in wealthy ones, not because of weaknesses with the ICT sector per se but because the broader environment is ill-suited to exploiting the technology.

Numerous authorities have discussed the growing digital divide. The ILO's World Employment Report for 2001 noted that "barely 6 per cent of the world's people have ever logged onto the Internet and 85 to 90 per cent of them are in the industrialized countries." Francisco Rodríguez and Ernest Wilson of the University of Maryland looked at rollout figures for a number of ICTs, asked "whether the data on ICT is characterized by *convergence* or *divergence* between developed and developing countries," and concluded that there is "a widening gap."[14] Bridges, a respected South African ICT think tank, concludes: "Real disparities exist in access to and use of information and communications technology (ICT) between countries (the 'international digital divide') and between groups within countries (the 'domestic digital divide'). . . . There is an overall trend of growing ICT disparities between and within countries."[15] Finally, the A. T. Kearney/*Foreign Policy* Magazine "Measuring

Globalization" study actually manages to put most of Europe on the wrong side of the digital divide, as well: "Rather than a division between developed and developing countries, however, the divide at this moment reflects the vast technological advances in North America and the Scandinavian countries compared with the rest of the world. Together, those two regions stand on one side of a gaping digital chasm that appears to have left much of the remaining world behind."[16] This claim has been repeated in numerous speeches and op-eds until it has the status of fact.

The case for a yawning digital divide *can* be made. In a recent paper that I wrote with a colleague, Carsten Fink, we found that, in high-income countries, when one adds together mobile and fixed connections, the telephone is so ubiquitous that there is more than one phone per person—compare that to a fixed and mobile teledensity of 3 telephones per 100 in low-income countries. The increase over the 1975–2000 period was by a little over 900 phones per 1,000 people in high-income countries, compared to just 24 phones per 1,000 people in poor countries. Looking at Internet use, an average of one-third of the population in high-income countries was using the technology in 2000, compared to just 0.4 percent in low-income countries. Given that effectively no one anywhere in 1990 had Internet access, it is clear that the absolute growth rate in terms of Internet access has also been far higher in rich countries than in poor ones over that period.[17]

But a large (even growing) *absolute* per capita access gap should come as no surprise. Telecommunications and Internet services are (indirectly) part of GDP—both from the expenditure side and from the income side. It is true that ICT expenditure accounts for a larger percentage of the economy in richer countries—low- and middle-income countries spend approximately 5 percent of GDP on ICTs, compared to 9 percent in rich countries.[18] It is also true that ICT access in developing countries tends to be more expensive,[19] due not least to large rural populations (59 percent of the population in low- and middle-income countries is rural, compared to 24 percent in high-income economies) with limited infrastructure access (about 2 billion people worldwide lack access to electricity, for example).[20] But the big reason is that rich country consumers have more money to spend than do poor country consumers.

In fact, finding no or a negative correlation between ICT stocks and GDP per capita would be anomalous, to say the least. If per capita expenditures on telecommunications were as high in the Philippines as in the United States, they would equal the former country's gross national product.[21] That would not leave much money to buy bread or water. Countries poorer than the Philippines (China, India and most of Africa, for example) would have to spend more than their gross national product each year to equal per capita US telecoms expenditures. This suggests that the idea of equal per capita stocks or use worldwide will have to wait upon far more equal distribution of income worldwide and explains why over 80 percent of the cross-country variation in stocks of telecommunications or Internet users at any one time can be explained by GDP per capita.[22]

Furthermore, a widening absolute gap in per capita ICT access does not necessarily imply that poor countries are falling behind. To assess this, we really should look at *relative* rates of growth. And the ICT gap between rich and poor countries is closing very fast in relative terms.

Since 1980, telephone penetration has been expanding faster in low-income countries, and considerably faster in middle-income countries, than it has in high-income countries.[23] Half of the world's households now have a fixed telephone. The rollout of mobile telephony—and particularly mobile phone *use*—has been phenomenal. There are more mobile than fixed phones worldwide today, considerably more mobile phones in developing countries than in the developed world, and the mobile footprint (the area in which people can make and receive a mobile phone call) covers 4.7 billion people—77 percent of the world's population. Survey evidence suggests that access to telephony is at least as high, with 88 percent of South Asians estimated to have access in their town or village, and 75–80 percent of people in Ghana, Uganda, and Bostwana having made a telephone call in the last three months at the time of the survey.[24]

Turning to the Internet, growth rates of users per capita have been higher in poor countries than in rich ones since the early 1990s—pretty much from the birth of the Internet, in other words. Remarkably, during the "Internet boom years" of the late 1990s, per capita usage was growing twice as fast in the developing world as it was in the developed world. The most stunning feature of the digital divide is not how large it is, but how rapidly it is closing. It is closing at a rate probably unprecedented for a communications technology. It took over 100 years for the telephone and close to fifty years for the television to reach a global "saturation level" of 10 per 100 population. For the mobile phone, it took just fifteen years, and it looks as if Internet use will equal that record.[25] There are more Internet users in the developing world today than in the developed world.

It is even possible to argue that developing countries are leapfrogging the developed world—seeing faster growth and a larger share of the world's ICTs than would be expected, especially given the lower percentage of GDP spent on ICTs in developing countries. Using ICTs per unit of GDP as a guide, Carsten Fink and I found a "digital leapfrog" in the case of telecommunications. Middle-income countries have the greatest numbers of telephone lines for each dollar of GDP, followed by low-income countries and high-income countries. And the leapfrog—at least between middle- and high-income countries—widened substantially in the late 1990s, in absolute terms. In the case of Internet usage, we also found middle-income countries surpassing high-income countries. Countries in the middle-income range surpassed the developed world somewhere in 1998 or 1999. Low-income countries still lagged behind high-income countries in 2000 but were clearly on a convergence course and had, by 2003, overtaken the developed world in terms of users per dollar of GDP.[26]

Turning to where Internet use is likely to have its greatest economic impact—in companies—there is evidence of an even greater catch-up, or leapfrogging, depending on how one wants to look at it, than there is with usage in general. The access gap is comparatively small between firms in LDCs and those in the West—much smaller than cross-country income gaps or national figures on the digital divide would indicate. For example, in 1999, approximately 21 percent of the population of high-income countries used the Internet, as compared to 2.2 percent in Eastern Europe and Central Asia. Yet Internet access in companies averaged around 60 percent in the Group of 7 (G7) countries, as compared to an average of 33 percent among firms surveyed in Eastern Europe and Central Asia. In other words, the Eastern Europe–G7 access gap for individuals was a factor of about ten, compared to an access gap of less than twofold for companies.[27]

Figures for Tanzania are even more stark. About 0.1 percent of the general population had access to the Internet in 1999, compared to 16 percent of enterprises—the Tanzania-G7 access gap for individuals was a factor of over 200, compared to an access gap of less than fourfold for companies.[28] And since then, the gap has shrunk at an incredible rate—a 2003 business survey found that 58 percent of businesses in Tanzania used e-mail for interacting with clients and suppliers. In Kenya, that same number reached 78 percent, and the average for fifty-two developing countries surveyed between 2001 and 2003 was over 50 percent.[29] Furthermore, it's not just the business digital divide that is smaller—94 percent of respondents to a recent survey of research institutes across the developing world said they had web and e-mail access.[30]

ICTs as Tools of Business

Despite lower investment as a percentage of GDP, access as a percentage of GDP is comparatively very high in developing countries, especially among businesses. Access is clearly of a different quality (computers are shared between more people, bandwidths are lower, and so on), but basic access is very high. Given that, there is surely the potential for considerable impact. And, indeed, there is evidence that significant growth in access rates, especially for businesses, occurs in large part because ICTs are very useful tools—especially for companies with overseas clients and expensive international call costs.

We have strong empirical examples of the older information and communication technology of telecommunications improving the functioning of markets. In 1967, Albert O. Hirschman of the Institute of Advanced Study at

Princeton offered evidence that a credit market for the coffee trade developed in Ethiopia only after the installation of a long distance telephone network.[31] Again, an ITU study of factories in rural Bangladesh found that the introduction of a telephone line reduced the amount of management travel and associated travel costs by a factor thirteen times the cost of installing the line.[32] There are numerous other studies on the positive impact of the telephone on the prices received for crops by LDC farmers and the creation of rural nonfarm businesses, employment, and other factors related to the quality of life.[33] There is even a growing body of macroeconomic and cross-country evidence linking telecommunications rollout with economic growth.[34] It is comparatively uncontroversial, then, that the rapid spread of telecommunications access has been a considerable force for improvement in the quality of life in developing countries since 1995. Given that the spread has been more dramatic in developing than developed countries, the telephone may have progressed from being a technology of divergence to one of convergence. (Although, given the dramatic nature of the rollout and continuing slow economic growth in a number of developing countries where that telecommunications rollout has occurred, the telephone is no panacea for slow growth.)

Turning to the Internet, one firm in Africa that adopted it early was Regent Clearing and Forwarding of Tanzania, an import-export company. It now uses ten cent e-mails to place orders rather than twenty dollar faxes and has reduced its monthly phone bills by 90 percent.[35] One estimate for Mozambique suggests that, including all costs, a company substituting e-mails for faxes to international clients for 10,000 pages a year would save over $6,000 a year.[36]

Regarding more advanced use, Caite Industrial Textiles of Brazil, which provides finishing for cotton and synthetic fabrics, began using a company called Viasebre to provide e-commerce services in 1999. The firm has automated service orders for 70–80 percent of its clients and has used the web to access financial information and monitor representatives. Almost immediately, the company saw its phone bills drop by 25 percent.[37]

Internet access can also drive sales. In Kenya, for example, the Naushad Trading Company, which sells local wood carvings, pottery, and baskets, saw revenue grow from US$10,000 to over US$2 million in the first two years the company went online.[38] More broadly, there is some cross-country evidence that developing countries with more Internet access are exporting more to the developed world.[39]

Indeed, access to ICTs is becoming a prerequisite for international businesses. A recent survey of international firms in Hong Kong, Singapore, and Taiwan found that the presence of advanced infrastructure, including ICT, was the most important consideration in the placement of regional headquarters, services, and sourcing operations. It was the second most important factor in determining production siting.[40]

The "Real" Digital Divide: Limited Utility of Advanced Internet Use

At the same time, it is important to note that actual usage rates for (rather than access to) the Internet in LDC companies are much lower than in the developed world. A survey of Tanzanian firms found that among the 30 percent of firms that had access to the Internet, less than half used it frequently, and only 13 percent rated it as an effective product promotion tool.[41] Although low-income countries have the share of global Internet users that would be expected given their income, they have considerably fewer hosts and broadband subscribers.[42] Of 110,498 secure servers worldwide that use encryption technologies in Internet transactions (commonly used for e-commerce), only 224, or 0.2 percent, are in low-income countries.

John Humphrey, a professor at the Institute of Development Studies (IDS) at Sussex University, carried out a survey of e-commerce use in developing countries with colleagues from IDS and the London School of Economics that confirms this impression. They looked at seventy-four firms in the garment and horticulture industries in Kenya, South Africa, and Bangladesh, all of which had access to the Internet. Of these firms, 66 percent had no website and 77 percent (and 90 percent of firms with over 500 employees) had not registered with an e-marketplace. Furthermore, only 7 percent had completed a sale as a result of listing on an e-marketplace. At the same time, it is worth noting that 95 percent of respondents in the garments sector were using e-mail to place or accept orders with existing clients. The Internet was useful to them, then, but for e-mail, not e-commerce. That suggests that the true digital divide is primarily related to advanced use. Similar results—suggesting computer access but little advanced usage—have been found in surveys of firms in Brazil, China, Malaysia, Mexico, and South Africa.[43]

When companies in developing countries are asked why they are not using the Internet or why they are not using it more intensively, they suggest that they see little potential benefit in advanced use. For example, the survey respondents in John Humphrey's study stated that the reason for not listing with e-marketplaces was that they saw little chance of increasing sales through such marketplaces.[44] In Chile, 37 percent of microenterprises have Internet access. Among medium-size enterprises, this figure rises to 85–93 percent. But even small enterprises don't see cost or ability to access as the major factor for avoiding use. Fifty-one percent of those firms suggest that they are not connected because they have no need to be.[45]

Furthermore, size is no predictor of a company in Chile using the Internet to sell goods and services online. Eighteen percent of microenterprises sell goods on the Net, and 18 percent of medium-size enterprises do the same. It appears that available investment resources (or the lack of same) have little to do with overall e-commerce levels in the country, and that conclusion is reinforced by responses to the question, "Why aren't you selling online?" in the Chilean survey. The two most popular answers are that it is not necessary (54 percent) and that clients don't use the Net to buy goods (23 percent).[46]

Similar findings emerge from a study of IT use among small and medium-size enterprises (SMEs) in East Africa, based on surveys in Kenya, Tanzania, and Uganda.[47] The percentage of tourism firms that have a computer in Uganda is five times, and in Tanzania ten times, the percentage of food firms equipped with a computer (10 percent of Ugandan tourism SMEs and 36 percent of Tanzanian tourism SMEs have a computer). This despite the fact that the average capital stock of a food enterprise is approximately twice that of a tourism enterprise.[48]

Across the developing world, survey evidence from over fifty countries suggests that 70 percent of firms in ICT services, real estate, and hotels and restaurants have websites compared to below 30 percent in agriculture, food, and auto components industries. The percentage of domestic microenterprises that do not export products and have Internet access is below 30 percent, compared to nearly 80 percent of domestic microenterprises that do export goods.[49]

What this variation suggests is that computer ownership and Internet use is determined at least as much by considerations of utility particular to the sector or firm as it is by capital availability. There are fewer computers in East African food companies than in East African tourism companies and fewer websites among developing country domestic auto parts manufacturers than ICT service exporters because the Internet is more useful for ICT and tourism companies.

If macroeconomic factors were driving the net utility of advanced Internet applications and the computer, one would expect little variation across sectors within the same economy. If lack of capital was the major barrier to expanded use among SMEs, one would expect lower use in firms with less capital. Instead, the wide variation in computer ownership suggests the unsurprising result that where computers and the Internet are very useful (in attracting global clients to tourism ventures), SMEs buy computers. Where they are not so useful (in the local food industry), SMEs do not buy computers. It should be noted that statistics for developed countries suggest precisely the same, unsurprising result—that sectors where the Internet is likely to be more useful see higher usage of advanced applications. For example, in the European Union in 2001, 40 percent of business services firms had their own website, compared to 24 percent of manufacturing firms.[50]

What is perhaps misjudged about the more strident calls for government intervention and subsidy and about the more extreme support, spoken or assumed, for the diversion of resources from other aid priorities to Internet usage programs is the assumption that entrepreneurs in developing countries are driven by ignorance to slow Internet adoption and are wasting precious opportunities. This assumption, based on perhaps simplistic models of technology diffusion, is itself somewhat too simple.[51]

As the evidence above suggests, Internet adoption has, in fact, been incredibly rapid in developing countries, especially among firms. In areas where developing country entrepreneurs see the most value in the technology—tourism, for example—the Internet is ubiquitous even in poorer countries. There is little evidence that the ignorant entrepreneur in Africa is missing her opportunity; instead, she's grasping that opportunity when it really exists. And, of course, the corollary to that is, if some entrepreneurs are not jumping on the Internet bandwagon, it may not be their ignorance that stops them. Instead, it may be a sober judgment of the potential returns on Internet use compared to other investments.

Drivers of Low Utility

Having said that, there clearly are widespread economic factors in developing countries that deter use, especially by domestic firms. For many companies, the low quality of infrastructure itself is one factor. Gloria Bampo, manager of a Community Learning Center funded by the US Agency for International Development (USAID) and providing Internet access in Cape Coast, Ghana, describes some of the problems that Ghana's telecenters face with telephone access. Shortly after an opening week that launched the Internet access program, she says:

> all the phone lines to Accra [Ghana's capital city, and the only place where there were Internet service providers (ISPs) in the country] broke down, and access to our ISPs became impossible. . . . On other days, when the telephone lines are working, connectivity becomes a problem because of erratic and slow connections. Just as interests in the services are increasing, we are faced with the problems caused by limited bandwidth in the country.[52]

Low quality is particularly reflected in the limited bandwidth connecting companies to the Web. Africa had 249 megabits per second (Mbps) of international Internet bandwidth in 1999. Compare that to 1,060 Mbps for Latin America and the Caribbean (the combined bandwidth between the two cities of Sao Paulo and Buenos Aires and New York is about the same as total international bandwidth for Africa), 6,469 for Asia/Pacific, or 45,459 for Europe—the bandwidth from London to New York alone is twenty-one times the size of the total international bandwidth in the African continent.[53] Such physical infrastructure constraints make robust low-bandwidth applications such as e-mail more effective than web-based transactions, for example.

Limited Internet use among locals is undoubtedly a second factor deterring use, especially in explaining differential use between local and exporting enterprises. If few customers are online, it makes little sense for domestic retail firms to be providing online services. And although the overall digital divide is closing rapidly, it will clearly remain very large in absolute terms. Because of significant income divides as well as the greater cost of access, it will be some time before the majority of Africans are frequently online, and those online today connect for shorter periods at service of a lower quality.[54]

But a larger constraint on LDC companies in adopting advanced uses of the Internet, especially for domestic purposes, is the broader environment to exploit the technology. Only 2.2 percent of India's Internet subscribers have engaged in e-commerce activities, for example.[55] Perhaps even more revealing are the statistics on e-commerce in Chile. It is one of the richest developing countries, with a government that has embraced the Internet and a range of advanced and popular government-to-citizen interfaces (42 percent of tax returns were delivered via the Internet in 2001, for example).[56] Clearly, Chileans have the capacity and access to use the Internet when they see an advantage to it. And yet, e-commerce sales in 2000 were worth 0.04 percent of GNP, and only 15 percent of the country's firms have their own website.[57]

This low level of advanced use also relates to a number of factors beyond infrastructure, such as the skills base of users and other supporting infrastructure, including the financial system. These factors create problems for enterprises attempting to put together e-commerce sites, but they will also reduce demand for such sites, if built.

Survey results from Botswana and Tanzania, for example, suggest that weaknesses across all types of infrastructure, combined with a low skills base—among employees and the population in general—and few relevant applications, deter advanced Internet usage in enterprises.[58] It is the general state of development in a country that deters widespread and advanced Internet use. The utility of any technology, especially a technology such as the Internet, depends on the broader environment, as we have seen. For the Internet, that utility is, as a rule, lower in developing than developed countries.

This involves, not least, a lack of access to human capital. A study of capacity building for electronic communication in Africa (CABECA), for example, found that 87 percent of Zimbabwean and 98 percent of Ethiopian Internet users had a university degree in 1998.[59] Yet, the stock of "tertiary human capital," or years of university education in the population, in LDCs, on average and as a percentage of US stocks, is about as small as the stock of physical capital.[60] Advanced education, of the greatest value in a global knowledge economy, is rare in LDCs. Indeed, approximately one-third of adults in low-income countries are officially illiterate.

Furthermore, even many of those officially classed as "literate" lack the level of reading skills necessary to fully utilize the Web. An OECD study carried out in 2000 found that although Chile had an official illiteracy rate of only 5 percent, only one-fifth of the people in the country had the reading skills to integrate sources of information and learn new skills from online sources. Chile is one of the richest developing countries, suggesting that this problem will be considerably more severe in regions such as Africa and South Asia.[61]

Looking more specifically at skills related to the Internet, the extent to which basic computer skills are lacking in LDCs is suggested by a report from the computer training center in Wa, in northern Ghana, that locals trained in computer skills and management could fetch $6,000 per year—in a country with an average gross national product (GNP) per capita of $390.[62] Further, these skills gaps are likely to remain in the population at large—not least because, with per student discretionary expenditures in secondary schools running as low as twelve dollars a year, the majority of schools in developing countries could not afford to install IT labs.[63]

Making this situation worse, skilled workers are the very ones who find it easiest to migrate to wealthy countries—already, some developing countries have lost as many as one-third of their skilled workers to migration.[64] It is not a complete loss, in that worldwide, foreign workers are estimated to remit $75 billion to their home countries each year. Nonetheless, it suggests that developing countries face a significant challenge in retaining expensively-nurtured human capital necessary to exploit the skills-based technology of advanced Internet use.[65] Given that the market in the United States for such workers is constricted—despite a much higher stock of human capital and the ability to import skills[66]—the skills-shortage in developing countries that are hemorrhaging their most talented IT professionals to the developed world must be even worse.

To see more widespread advanced use of the Internet, a second specific type of human capital is required: language skills. Language remains a significant barrier to use, as suggested by a study of users in Slovenia, which found that 75 percent of those who considered themselves fluent in English used the Internet compared to 1 percent of non-English speakers.[67] Results from a study conducted in Tokyo, Beijing, Seoul, Bangkok, Singapore, and Jakarta suggest that English speakers were two to four times more likely to use the Internet than the non-English-speaking population.[68] More generally, across countries and allowing for a range of other factors, countries where English is the official or most widely spoken language see significantly higher Internet users per capita.[69] That is hardly surprising, given the quality and quantity of non-English material on the Web. In 1999, 72 percent of websites were in English (although that proportion has since fallen). And there is a significant language skills gap, especially among poor people in LDCs, with perhaps one-half of the populations of the least developed countries not speaking an official language of their own country—let alone English, the predominant language of the Internet.

Beyond the scarcity of physical and human capital needed to benefit from the development of many advanced Internet applications, the institutional environment in LDCs is inconducive to rapid and successful exploitation of the technology. For example, online markets face significant "information asymmetry" problems, which are most severe in auction markets, where traders are identified only by e-mail addresses. That has already led to some significant cases of online fraud and, in turn, greatly increases the importance of consumer protection laws—the right for buyers to withhold payment until there has been an investigation of fraud, for example.

Weak institutional capacity has been found to correlate across countries with lower access to networks and lower host site development.[70] Weak institutions also lower consumer trust in e-commerce, perhaps the most important factor in determining willingness to purchase online.[71] One recent study of the distribution of host sites around the world found that, beyond access to computers and phones and costs of telephony, a measure of the rule of law (as well as a measure of credit card use) was a strong determining factor. Thus stronger legal environments are likely to lead to more e-commerce.[72]

Poorly developed financial systems in particular, especially when combined with poor physical communications infrastructure, can significantly reduce the potential for e-commerce in LDCs. For example, a recent survey of business trust in the postal service found that willingness to entrust the postal network with a package worth $100 was strongly correlated with GNP per capita, with Finland, Japan, and Switzerland at the top and Venezuela, Honduras, and Nigeria at the bottom.[73] Regarding credit cards, results from Latin America suggest that only 28 percent of online transactions in the region involve credit cards, compared to 54 percent using cash—and that has more to do with lack of trust in than lack of access to the credit card system.[74] In China, only 13 percent of online transactions are completed with online payment, compared to 42 percent cash on delivery and 24 percent remittance via the post office.[75]

A number of "soft" institutional factors make the successful launch of complex Internet-based ventures even more difficult in developing country settings. Richard Heeks of Manchester University—who has long studied the use of IT in developing countries—lists a number of such factors: a lower valuation of formal knowledge, contingent work processes, cultural barriers, and hierarchical management structures.[76]

Mike Dertouzos, at the MIT Media Lab, found his optimism for the impact of the Internet in developing countries tempered by such constraints that he encountered in Nepal:

> A few of us techies got together with a colleague from Nepal, fully expecting to boost his nation's economy by 20 percent through clever use of the information marketplace. Unfortunately, we quickly found out that even if we

> got the communications, hardware, software and training for free, we would still fall short of our goal: Only 27 percent of the Nepalese are literate. And of these, only a small fraction speak English. When we asked what services that smaller group could offer, we hit a brick wall. Many are not skilled, and those who are already are busy running their nation's businesses. Maybe we were too ambitious when we envisioned a future workforce in Nepal selling office services to New York and London via the Web. What if we focused instead on selling Nepal's famous crafts, like custom-made rugs, on the Web? That got us into all sorts of other concerns about establishing trust among distant buyers and distributing the goods. The potential of the modern information age seemed overshadowed at every turn by the ancient forces that separate the rich from the poor.[77]

It is likely that these factors doomed a number of ventures based on advanced use of the Internet in LDCs.

There are examples of successful use of e-commerce in low-income environments, such as the Naushad Trading Company in Kenya, mentioned above. On a larger scale, a network of Indian telecenters called e-Choupals use networked computing applications to provide information as well as sales services to farmers in rural Madhya Pradesh, a state in central India. ITC, a company that sells farm inputs and buys crops from farmers, set up an intranet utilizing phone and very small aperture terminal (VSAT) connections that serves as many as 1 million farmers through an intermediary operator. Each e-Choupal costs between $3,000 and $6,000 to set up but reduces ITC's procurement costs by about 2.5 percent while raising prices paid to farmers by about the same amount. E-commerce can work even in rural areas of developing countries then. But ITC's is a fairly unique case where a 5 percent improvement in efficiency was large enough to sustain a computer network, not least because ITC has such a significant market share of both agricultural inputs and crop purchases, because the agricultural market is comparatively advanced for a low-income country (comparatively large farmers growing soy using intensive techniques), and because the sites chosen for e-Choupals tended to be near major cities with good road and telecommunications connectivity.[78]

That the set of circumstances that has apparently allowed ITC to roll out e-commerce at a profit in a low-income country is fairly rare might help to explain why a recent survey of e-commerce use in least developed countries was able to find only a few examples of e-commerce, largely serving niche markets, limited to sales between $2,000 and $30,000 per year and employing a maximum of fifty people—similar to the example of Naushad when it started out.[79] Peoplink.com provides an example of how difficult success can be, even when backed by a range of donors. It was an early program designed to showcase the business benefits of ICTs in a developing country setting. Third World artisans were able to sell their goods over the Web through Peoplink's website, and 16,500 signed up. The site got glowing reviews from the *New York Times* and *Newsweek,* among others. But a study of those producers found little evidence that Peoplink had increased sales, and (perhaps as a result) the marketing site is no longer active.[80] Again, Viasebre, mentioned earlier as the provider of business-to-consumer e-commerce services in Brazil, cost about $750,000 to set up, had monthly running costs of $17,000, but was making only in the region of $7,000 per month in 2001.[81] More broadly, a survey of ninety-two South African businesses already involved in e-commerce found that 84 percent did not see any increase in revenues as a result of moving online.[82]

We have seen that the digital divide has shrunk remarkably rapidly. It is now no larger than would be expected, given the income divide between countries. When it comes to businesses, the digital divide is already considerably smaller than that. And where it still exists, it may well reflect an accurate calculation on behalf of individual company managers regarding the utility of the Internet for their business, rather than ignorance or inability to access the technology. If there is a digital divide to be overcome, it involves that larger calculation of utility.

And the trouble with the long list of larger factors (infrastructure, education, institutions) that create disincentives both to general business development and the use of advanced Internet applications in particular is that they are tightly linked with the broader level of development. In other words, the significant digital divide is the usual development divide—some countries are rich, and others are poor. Not surprisingly for a general purpose technology such as the Internet, its utility depends on the general level of development. In short, the best way to use the Internet for development is to use development to increase the utility of the Internet. That conundrum suggests the technology is an unlikely source of leapfrogging growth.

Surveys of firms in developing countries confirm the impression that information and communications infrastructure ranks far down the list of concerns for the average company, below some of the broader issues listed as barriers to advanced Internet usage above. Business survey evidence from over fifty developing countries suggests that communications networks rank last out of fourteen constraints to business growth, after factors including finance, skills, and access to electricity. Compared to 40 percent for the top-ranked concerns covering taxes, the macroeconomy, and policy uncertainty, less than 10 percent of firms rank communications networks as a major constraint on doing business (and the number drops to about 5 percent in countries that have undertaken telecommunications reform).[83] Lack of access to the Internet does not appear to be a significant barrier to development. However, general development is a significant barrier to advanced Internet use.

Thus it is unsurprising that the limited amount of direct analysis we have on returns on ICT investment in developing countries suggests an even more tenuous connection with productivity than OECD studies. At the microlevel, a

1996 study of manufacturing firms in northern India found that IT investments were correlated with more highly skilled workforces, but not with wages or productivity, as we have seen.[84] A second study of IT use among SMEs in East Africa suggests that the proportion of a firm's capital dedicated to ICT has no significant impact on the internal rate of return and is associated with *lower* labor productivity.[85] Meanwhile, even cross-country studies that find a historical link between IT investment and growth in developed countries fail to find such a link in LDCs.[86] Despite increasing imports of high-technology goods (especially computers) over the last few years, there is little evidence of improved labor productivity in many low-income countries.[87] Although the late 1990s saw ICT capital deepening play a greater role in the Asian growth story, overall TFP growth dramatically declined, suggesting once again a muted impact of the ICT revolution.[88]

Echoing these earlier results, a recent study of ICT and economic growth has found that the link between ICT penetration and economic growth is weaker in developing countries than in developed ones. The same study found no evidence of a statistically significant and robust relationship between investment in ICT and TFP growth across countries. (The study also found that countries with more effective legal systems, more educated populations, and English as the major language saw greater benefit from ICT investment, which is not surprising, given earlier findings on language as a barrier to Internet use).[89]

The Internet as a Force for Divergence?

Indeed, given the above, there are some reasons to believe that the Internet might be a force for divergence. It is the underlying feature of the exogenous growth story that we discussed in Chapter 1 that technology flows across borders with relative ease. Endogenous growth theory, which suggests the importance of physical and human capital in which technology is "embodied" and stresses the role of institutions, is less sanguine about the impact of the average new technology on developing countries catching up to developed countries.

If the history of older ICTs supports one interpretation of the development story over the other, it is endogenous theory that comes out on top. As we have seen, Tony Venables of the London School of Economics suggests that innovations in communications technologies in the past have further concentrated income in a few geographic areas.[90] With the appearance of the telegraph, for example, the result was not greater independence of overseas partners of transnational firms but absorption by the "home" office.[91] Again, the transportation revolutions of the nineteenth century did not lead to the dispersion of economic activity but instead to its concentration, as the value of proximity to consumers was reduced in comparison to the benefits of agglomeration externalities and increasing returns to scale.

Distance has yet to die. At the firm level, ICT-supported innovations such as just-in-time management actually increase the cost of physical distance. Perhaps for that reason, since 1985, the average distance between supplier plants and assembly plants in the US auto industry has actually fallen, as we saw in Chapter 3. And it appears that the Internet is further concentrating supply chains among a technological elite.

Why? First, it turns out that most things that can be done from a distance using ICTs are better done in wealthy countries. We saw in Chapter 3 that most of the services that are now contestable by developing world companies—such as airline ticketing—saw a dramatic replacement of labor by technology prior to that contestability, reducing the benefit of attracting such industries to an LDC. Meanwhile, developed world firms can now better compete with local developing country providers of services such as banking and consulting. Some further evidence that e-commerce might be eroding local advantage in production in developing countries is that, in the United States, approximately 9 percent of online purchases are made from overseas—compared to 75 percent in Chile.[92]

To balance these fears, however, it appears to date that the Internet's ability to determine which companies in the developing world get business has been limited. John Humphrey and his colleagues' survey of e-commerce in developing countries finds that the major effect of B2B e-commerce is to strengthen the relationships between existing trading partners. At least in their surveyed companies (all had access to some level of Internet connectivity), LDC firms were not losing market share as a result of the spread of the Internet.[93]

Nonetheless, reasons for concern may remain. Economic concentration as a result of improved communications might also be related to the different mix of exports and imports produced and consumed by developed and developing countries. The few developing country companies selling differentiated labor-intensive products (including software or remote processing) might see an increase in demand if they can prove themselves more efficient at production. But such companies are the exception in developing countries, where most exports are commodities or "commodified" manufactures.

And even though commodity trading is likely to become a little more efficient, it is also likely that the savings accrued will be appropriated by consumers rather than producers. Though we have seen that the evidence is not as strong as sometimes assumed, the Internet is a better tool for allowing price and product comparisons in the market for homogeneous goods and services (raw materials and standard manufactured items that are easily compared on a cost-only basis) than for heterogeneous goods (word processing systems, for example), where quality and other factors have to be taken into account. It is the market for homogeneous goods that is likely to become increasingly competitive because of the Internet. And as homogeneous, commodified goods make up the majority of exports in many of the poorest countries in the world, it is possible that one impact of the Internet might be to worsen the terms of trade faced by these countries.

Neoclassical trade theory points out that global welfare will be improved by this change. Rich country consumers will be better off, as will some more efficient producers in developing countries. However, as benefits will be concentrated in developed countries and among a small number of developing country producers, the overall impact will still be to speed divergence of incomes between countries. Indeed, even if the benefits of cheaper transportation costs and Internet-enabled production management allow for some goods to be produced more effectively in low-cost environments and even if only weak agglomeration effects remain, Venables points out that it is likely that only a few areas will benefit from increased ICT-related economic activity in the Third World. All this, he notes, fits the pattern as regards the Internet so far, which has seen outsourcing activities concentrated in India. It is also true of LDC export industries for telecommunications equipment, which have expanded rapidly as a share of total equipment trade but remain concentrated in four or five LDCs.[94]

Certainly, the increasing reach of patent laws means that technology transfer comes with a growing price. The debate over AIDS drugs in the Third World suggests an obvious truth—poor developing countries get more out of technologies when they are "commons" technologies than when they are wrapped around with patents and copyrights that make them far more expensive to use. The technology that runs the World Wide Web (TCP/IP—Transmission Control Protocol/Internet Protocol) is a commons technology, and over 50 percent of software is currently pirated in the Third World (with rates reaching as high as 97 percent in Vietnam).[95] But content on the Web is being controlled by new laws designed to give great power to its creators worldwide. The US Digital Millennium Copyright Act, for example, bans technologies designed to circumvent copyright protection mechanisms and has been used to arrest a Russian programmer visiting the United States who (in Russia, where it is legal) wrote code to allow e-Books to be moved from one machine to another.

And patenting has spread to many "products" that differ from the Alexander-Bell-researched-and-built-it-in-his-laboratory traditional image (indeed, to products that look a good deal more like institutions). That is particularly true of the Internet, and the US Patent Office has granted patents on nonmaterial processes, including group buying, one-click shopping, and reverse auctions. This US government–enforced monopolizing of "disembodied" technologies has international ramifications, because the World Trade Organization (WTO) agreement on Trade-Related Aspects of Intellectual Property Rights (TRIPS) forces LDC members to accept minimum standards on patents, copyrights, and trademarks, based on OECD practice. As patents become more widely applied across types of inventions and around the globe, the ability to catch up in (and benefit from) the race to produce goods enabled by ICTs becomes more complex.

International recognition of copyrights and business applications patents ensure that most benefits of technological advance go to inventors rather than consumers and reduce spillovers as a proportion of new technology, even while they might (arguably) promote the creation of such technology. If that is the case, it is likely that these benefits will increase divergence between technology-producing countries and others.

Most important, perhaps, the exogenous concept of "free" technology, central to the standard growth story, may be completely reversed in the context of ICTs, with negative network externalities forcing inappropriate and expensive technologies on the developing world (such as word processing applications in which only a small minority of commands are used).

Because of those trends, updating software, even when it is pirated, still carries significant costs for the developing world. Software updating demands hardware updating, and hardware can't be pirated. Software updating also requires technical training, and human capital cannot be pirated, either. Robert Wade of the London School of Economics argues that there is a "software-hardware arms race" going on worldwide—the global version of the baroque network problem discussed in Chapter 3. Computer manufacturers build ever-faster computers to run new versions of software with large demands on processing power, and software manufacturers design ever-more baroque programs to exploit the new power.

As Wade argues, it is doubtful that this process of upgrading is being driven by value added, even in developed countries. Instead, he argues that managers at large companies adopt the new software at the behest of IT staff more concerned with maximizing their budgets than with company profitability and out of concern that the "brand" of the company will suffer if it is found using old stocks of IT. But more to the point, Wade notes that constant updating affects other companies in the developed and developing worlds, who have to buy new computers in order to keep up—to be able to exchange documents, for example.

At the same time that patent laws and network economies garner high returns on technical innovation in the United States, technological change spurs depreciation of capital worldwide. For example, the collapse of the Russian economy could be characterized as the impact of formerly excluded Western technological change "depreciating to zero" a large part of Russia's human and physical capital stock. The rapid pace of US-led technological change in computing is doing the same to the value of older computers worldwide. It is not just an accounting issue. For IT in general, much of the value of a particular piece of equipment lies in its being compatible with the broad stock of IT. The value of a computer with 1990-style characteristics is much less now than it was then—then, it could swap large floppy disks with other computers and display the latest version of Word documents produced on other machines.

Now it would be difficult to get the software and even harder to share documents. The "network externality" element of a 1990 computer has been completely removed by technological progress. Because the broad stock of IT resides in the developed world, rich countries determine the pace of global technological depreciation. As a general rule, then (with many and obvious exceptions), the value of LDC IT will be reduced if it is not compatible with industrial world IT.

Even technologically (more) appropriate IT solutions for the developing world, such as the simputer (the $300 Indian-designed computer), face this problem. Even though it runs on nonproprietary software, if the simputer doesn't allow rural Indian users to receive e-mails from their relatives in the UK—because of proprietary software needed to look at attachments that the Indian users don't own, or because of multi-megabit headers—its value is diminished. If it doesn't allow rural Indian users to e-mail their relatives in the city who have upgraded to ensure they can e-mail their relatives in the UK, its value is still diminished. That may help to explain why only 4,000 Simputers were sold in 2004 and 2005.[96]

But there isn't much that India can do about that. It controls too small a stock of global IT to ensure that all applications remain "simputer compatible." During fiscal year 1997, Microsoft spent $1.93 billion on product research and development. That is about 40 percent of total R&D expenditure in all low-income countries worldwide that year.[97] Industrialized countries hold 97 percent of patents worldwide, and half of global licensing and royalties fees were paid to just one country—the United States. In 1997, the United States was home to five of the six top global personal computer manufacturers by revenue, the three top operating system manufacturers, the three biggest browser makers, eight of the top twelve telecommunications companies, the two biggest Internet service providers, the three largest search engines, and three of the five largest information and entertainment companies.[98]

Microsoft is a perhaps particularly significant part of the problem related to IT patenting. Its control of the operating system gives it a good deal of say over all the software developed for the system, and software development also further establishes the system as economically dominant. Creating a suite of programs "given away" with the operating system further entrenches the system, its standards, and Microsoft's ability to issue new versions that become must-buys. In the browser war with Netscape, for example, Microsoft gave away copies of its Explorer program because, as Bill Gates said, "we don't need to make any revenue from Internet software." Thus it used the program to reinforce its overall technology and operating software dominance.[99] I do not mean to say that Microsoft is doing anything particularly immoral—it is merely acting to maximize its profits, taking advantage of network externalities and market distortions created by intellectual property rights. That is what firms are meant to do. Nonetheless, the result of this legitimate effort to increase profits in a broken marketplace is causing economic harm—perhaps especially to developing countries.

It might be, then, that US IT companies such as Microsoft are reaping profits at the expense of the wider US market (perhaps helping to account for some of the huge returns seen in the IT sector, compared to limited effects on the economy as a whole), and the United States as a whole is making a profit at the expense of the rest of the world (which, in turn, might help to explain the limited impact of IT investment in Europe and LDCs). In 2000, US companies produced 56 percent of the revenues and received 96 percent of the profits in the global IT production business. IBM, Hewlett-Packard, and Dell alone delivered 96 percent of net global profit in the computer manufacturing business. US total profits in computing were 130 percent of net global profits—the average computer firm outside the United States lost money. Microsoft, Oracle, and Cisco alone accounted for 69 percent of global net profits in software production.[100]

One potential way the Web may support convergence is by allowing scam artists in developing countries with weak enforcement of antifraud laws to make money off wealthy-country victims. The Caribbean has its 809 area code fraud: an e-mail or voicemail suggests you call a number starting with that code because you have won a wonderful prize or your sister has been returned by aliens, and you are charged multiple dollars per minute to listen to a recorded message until you eventually hang up. Nigeria has the "strictly confidential" e-mail from Mr. Oluwa about a bank account with $300 million in it, which he's willing to share with you if you just give him your bank codes. Apparently Nigerian-based e-mail fraud like this actually works—in 2004, sixteen Americans claimed financial losses totaling $345,000 from the scam, quite possibly a fraction of the total amount lost due to this fraud.[101] Nonetheless, it appears that the Internet generally pushes toward divergence in incomes.

Neoclassical trade theory seems a little weak as a predictive science. Having said that, even neoclassical trade theory provides some reason to believe that the Internet might be behind the growing divergence of incomes between rich and poor. Beyond that, the hard evidence from prior "information revolutions" and the evidence to date regarding the Internet suggest that divergence is a plausible outcome. Perhaps the only positive note is that, much as the overall impact of the Internet on the global economy may have been overplayed, any significant divergence as a result of the technology must also be unlikely.

Estimating the Overall Economic Impact of the Internet in Developing Countries

Based on the discussion of the rapid rollout of ICTs in the developing world but also on the limits to advanced use created by a broader environment inconducive to a networked economy, is it possible to estimate the size of the impact of the Internet on developing country economies. Not with any degree of certainty, but there are some optimistic forecasts of the greater macroeconomic efficiency that will be forthcoming from Internet usage in LDCs, particularly in e-commerce. Ignoring any potential divergent impact, then, what is the largest plausible positive impact that we can expect from the new technology?

It is too early with the Internet in particular to even attempt the somewhat questionable cross-country regression analysis that suggests that there might have been a link between telecommunications rollout and economic growth across countries since the 1960s or so.[102] Regarding sectoral TFP analysis of the type we used to examine the impact of ICTs in the United States and the OECD, such data are only available for a few developing countries.

Nonetheless, we can make some educated guesses. Probably the biggest economic impact of the Internet is to lower business transactions costs. Given the likely scale of those transactions cost changes, compared to previous "revolutions" in transactions costs, the picture, yet again, looks somewhat grim. Assuming generously that the Internet might reduce transactions costs on goods and services by perhaps 10 percent, compare this marginal improvement to the scale of changes in "institutional technology" since the 1950s.[103] The impact of the General Agreement on Tariffs and Trade (GATT) and WTO trade rounds, for example, was to lower average tariffs in industrial countries from 40 percent in 1947 to 5 percent in 1988—and they have fallen further since.[104] Falling transportation costs and more rapid transportation systems might have added a further 10 percent cost reduction on top of that.[105]

But despite (or perhaps because of) such dramatic reductions in the cost of international commerce since the 1950s, LDC income growth rates have fallen over that period, even further behind those of developing countries, suggesting that reduced transactions costs alone will not allow LDCs to "close the gap." Indeed, cross-country studies have a difficult time showing that greater openness to trade is connected with growth outcomes at all.[106]

The growth estimates that have been produced, even by optimists, reflect the reality that e-commerce will likely be a marginal motor for growth in developed and developing countries alike. A range of estimates for OECD countries suggests an impact of e-business on growth as small as perhaps one-third of 1 percent in the period up to 2005. Taking the United States alone, a more generous estimate by Goldman Sachs is perhaps 5 percentage points of GDP impact by 2010.[107] More optimistic still, it has become common to assume that investments in telecommunications and IT, which account for the same proportion of today's capital stock as railways did in the late nineteenth century (a little over 10 percent), will have a similar impact on US economic growth as is generously estimated for the railways—around 10 percent (although Robert Fogel would estimate one-third of that impact for railways). If that were gained over the next twenty years, it would be at a rate of a little under 0.5 percent per year.

Even Internet optimists widely believe that the impact of the Internet on developing countries, at least over the near term, will turn out to be smaller than that in developed countries—and this chapter has presented a range of evidence suggesting that the impact will remain muted. The consulting firm e-Marketer estimated that e-commerce revenues of "the Rest of the World" (outside North America, Europe, and East Asia) would be 2 percent of the global total in 2003—which equals $US29 billion. Assuming that the impact on GDP is one-tenth that revenue, this suggests that e-commerce added the equivalent of Guyana's GDP (a poor country with a population of a little under 1 million) to total "Rest of World" GDP in 2003. Another global estimate suggests that, over the longer term, "effective" e-commerce policies could increase Latin America's GDP by about $45 billion—or about 2 percent. However, the same source provides other estimates that are as low as 1.2 percent for Latin America and 1 percent for Asia.[108]

These general estimates of the income impact of the Internet are thus very small compared to the rich-poor gap. The US GNP per capita is about $30,000, compared to Sudan's GNP per capita of about $300, or about a 10,000 percent difference.[109] Compare the latter figure to the 10 percent additional income that the Internet might most optimistically provide.

In short, ICT rollout in the developing world has been in many ways even more impressive than rollout in OECD countries. Mobile telephony in particular is likely to have a significant impact in poorer countries. The Internet has also found an important role as a tool for communications worldwide. At the same time, evidence that the Internet is an economically transformative technology is even more sparse in the developing world than in the developed, and there are good reasons to believe some Internet investments will produce very low returns. But even ignoring those fears, optimistic forecasts of the direct economic impact of the Internet through e-commerce still suggest an impact that is marginal, given the scale of the development divide between rich and poor. We will see in Chapter 4 that the impact of the Internet on delivery of government services, while potentially significant, is also likely to be marginal rather than revolutionary.

Notes

1. "How a Need for Speed Turned Guadalajara into a High-Tech Hub," *Wall Street Journal*, March 2, 2000, p. A1.
2. Kumar and Joseph 2005.
3. D'Costa 2003.
4. Qiang, Pitt, and Ayers 2003.
5. Ibid.
6. ITU 2000.
7. Qiang, Pitt, and Ayers 2003.

8. Heeks and Kenny 2001.
9. Calculated from World Bank 2001a.
10. APEC 2001.
11. Calculated from Joseph 2002.
12. D'Costa 2003.
13. Heeks and Nicholson 2002.
14. Rodríguez and Wilson 2000.
15. See http://www.bridges.org/spanning/report.html.
16. *Foreign Policy*, January 2001. And, in the interests of full disclosure, I was on the writing team of a report saying much the same thing. To whit: "The gap in [ICT] provision is large—much larger than income disparities for some regions. In particular, the gap is growing in provision of advanced services" (World Bank 2000b). In (only) partial defense, the paper did note that "some of the trends in ICT provision around the world have been toward convergence."
17. Data from World Bank 2002a.
18. Data from World Bank 2005. Carsten Fink and I found an approximately proportional relationship between (fixed plus mobile) teledensity and per capita income. For 2000, the elasticity of per capita telephone access with respect to income takes a value of 0.985 (estimated across 162 countries by a simple log linear regression of teledensity on an intercept and income). By contrast, we did find an overproportional relationship between per capita Internet usage and per capita income, with an estimated elasticity of 1.113 (estimated across 166 countries). Telecommunications investment alone is a higher percentage of GDP in developing than developed countries (World Bank 2005), suggesting, perhaps, that investors see higher returns on telecoms than on computing.
19. In 2000, hourly Internet access in Africa was estimated to cost $14, compared to $1.45 in the US. Fanda 2002.
20. Albouy 1999.
21. Calculated from ITU 2002 and World Bank 2002a.
22. Forestier, Grace, and Kenny 2001.
23. Other measures of convergence suggest the same thing: in a sample of 101 countries for the 1960–2000 period, mainlines per capita have seen the coefficient of variation drop from 1.76 to 0.98 over that time, the population-weighted coefficient of variation fall from 1.80 to 1.34, and the number of telephone lines in the bottom 20 percent of countries in terms of rollout expressed as a proportion of telephone lines in the top 20 percent of countries rise from 0.003 to 0.008 (see Kenny 2004).
24. Keremane and Kenny 2005.
25. Kenny, Lanvin, and Lewin 2003.
26. In 1996 the elasticity of per income Internet usage with respect to per capita GDP took a value of 0.567 (estimated across 118 countries by a log linear regression of Internet users over GDP on an intercept and per capita GDP)—suggesting that higher incomes still had a sizable association with greater per income use of the Internet. By 2000, this elasticity had fallen to 0.151, pointing to a substantial weakening of this relationship. The United Nations Conference on Trade and Development put together a measure of global gini coefficients for access to various ICTs over the period 1995–2001, which point in the same direction. Gini coefficient analysis is one way to measure inequality of ownership or access—it is a score from zero (everyone has an equal share) to one (one person has all of it). From 1995 to 2001 is a very short length of time, and it is likely that the gini coefficient of global income has probably changed hardly at all over that period (it has varied by less than 0.05 over the entire period from 1960 to 1997). Compare that to the results they found for ICTs—a very rapid decline

in the gini for Internet users from 0.87 to 0.73 and for mobile subscribers from 0.82 to 0.66—in just six years. Indeed, these latest figures make ICT inequality less severe than global income inequality measured at standard exchange rates (although still worse than those measured at purchasing power parity [PPP]). Nonetheless, the authors still concluded from their study that this was "evolution over time representing small, incremental reductions"—would that we could see such small, incremental reductions in global income inequality (see UNCTAD 2003b). Inequality data (using non-PPP exchange rates) come from Melchor 2001. It should be noted that the statement on long-term trends uses PPP exchange rates.

27. These figures are calculated from data in Clarke 2001, who gives results for Internet access in enterprises in the ECA region; the World Bank 2001a, which gives figures on Internet users and population for ECA and high-income countries; and the UK Department for Trade and Industry 2000, which gives data on Internet access in enterprises in the G7. The figure for enterprise access in the G7 is an unweighted average of access data at the country level.

28. Calculated from Pigato 2001 and World Bank 2001a.

29. Qiang, Clarke, and Halewood 2005.

30. See http://www.gdnet.org/survey2.htm.

31. Hirschman 1967.

32. ITU 1998b.

33. Duncombe and Heeks 1999, ITU 1998a, 1998b.

34. See Forestier, Grace, and Kenny 2001 for a review.

35. Africa Internet Forum 1999.

36. Ibid.

37. Heckl and Waack 2001.

38. Grace, Kenny, and Qiang 2004.

39. Clark and Wallsten 2004.

40. Mody 1997.

41. Pigato 2001.

42. World Bank 2005.

43. Molla 2005.

44. Humphrey et al. 2003.

45. Government of Chile 2002.

46. Ibid.

47. Choudhury and Wolf 2002.

48. This does not reflect differential access to capital; the average capital stock of a food enterprise is twice that of a tourism enterprise. Similar variations occur around the world—in India, 27 percent of plants in the motion picture industry have the Internet, as compared to none in sanitation (see Joseph 2002).

49. Qiang, Clarke, and Halewood 2005.

50. Eurostat 2002.

51. See Geroski 1999 for a review.

52. Bampo 2001.

53. Calculated from Abrahmson 2000.

54. Pigato 2001.

55. Miller 2001.

56. Economist Intelligence Unit 2002.

57. Hilbert 2001.

58. Pigato 2001.

59. CABECA 1998.

60. Heeks and Kenny 2001.
61. OECD 2000b.
62. Hirsch 1998.
63. Grace and Kenny 2003.
64. World Bank 1999.
65. Castles 2000. HIV/AIDs is making this situation even worse, especially in Africa. The disease disproportionately affects skilled, professional and managerial workers (see ILO 2001).
66. Perkins and Perkins 1999. The NRC report also notes that these workers have seen their real wages (excluding stock options and equity stakes) rise between 3.8 and 4.5 percent annually over the past three years, compared to 3.2 percent for other professional specialty occupations, suggesting that the IT labor market remains tight, despite the raising of the visa quota on IT-skilled immigrants in 1998.
67. Kenny 2002b.
68. Kenny 2002a.
69. Guillen and Suarez 2001.
70. Kenny 2003; Oxley and Yeung 2000. In a related finding, Guillen and Suarez (2001) discovered that a democracy index is significantly correlated with Internet users and hosts per capita, after allowing for a range of other factors.
71. World Bank 2004.
72. Oxley and Yeung 2000.
73. Kirkman et al. 2002.
74. Hilbert 2001. Hekl and Waack (2001) report that only 14 percent of Brazilian credit card owners consider using cards over the Net as somewhat or very safe.
75. Yusuf 2004.
76. Heeks 2002a.
77. Dertouzos 1999.
78. Kumar 2004; Annamalai and Rao 2003.
79. UNCTAD 2001.
80. UNDP Evaluation Office 2001.
81. Hekl and Waack 2001.
82. Molla 2005.
83. World Bank 2005.
84. Lal 1996.
85. Choudhury and Wolf 2002.
86. See Pohjola 2001; Mayer 2000. Bedi's 1999 survey on this topic could find only one study with a significant result, which found a positive relationship that was likely to be plagued by an endogeneity bias.
87. Mayer 2000.
88. Lee and Khatri 2003.
89. Khoung 2004.
90. Venables 2001.
91. Brown and Duguid 2000.
92. Hilbert 2001.
93. Humphrey et al. 2003.
94. See ITU 2002.
95. Business Software Alliance, 2000.
96. Farivar 2005.
97. Calculated from World Bank 2001a and Newman 1997.
98. Hamelink 1999.

99. Newman 1997.

100. Strassman 2000.

101. B. Koerner, "The Nigerian Nightmare," *Slate*, posted Tuesday, October 22, 2005.

102. Improved access to telephony has been found to correlate with faster economic growth in numerous cross-country econometric studies—for what they are worth. As part of a review with my colleagues Emmanuel Forestier and Jeremy Grace, I compiled a list of fifteen such studies, and I missed more than a few (see Forestier, Grace, and Kenny 2001). These studies looked at the statistical strength of the relationship between telephone rollout and economic growth and found it to be reasonably strong. It should be noted that, for good reason, quality statisticians point out that correlation does not prove causation and that all such studies can do is show that there is not statistical evidence *against* a suggested relationship. Further, a number of the study results fell apart when subjected to harsher statistical testing (Munnell 1992, Garcia-Mila and McGuire 1992). Overall, despite sharing the networking and transactions-reducing features of the Internet and despite human and physical capital requirements far less demanding than the Internet, it appears that telecommunications has had a fairly limited growth impact. A similar story can be told for the railroad. Fogel's 1964 study of the impact of the railroad on the US economy (the founding work of cliometrics) estimates that the level of per capita income reached in the United States on January 1, 1890, would have been reached by March 31, 1890, if railroads had never been invented (see Fogel 1964).

103. Goldman Sachs 2000.

104. Law Journal Extra 1996.

105. Hummels 1999, 2001.

106. See Edwards 1993; Maurer 1994; Rodriguez and Rodrik 1999; Walde and Wood 1999.

107. "E-commerce; The A to Z of B to B," *Economist* April 1, 2000. See also the *Economist* Survey: "The New Economy," September 23, 2000; World Bank 2000c; and OECD 2000a.

108. Mann, Eckert, and Knight 2000.

109. World Bank 2002a.

5

Experiments with E-Government in Developing Countries

GOVERNMENTS IN DEVELOPING COUNTRIES are frequently accused of being inefficient and corrupt. Ill-equipped civil servants operating with little public oversight deliver poor-quality services that frequently bypass the poor. For proponents of e-government, the Internet is well designed to overcome these problems. It is a powerful tool of information transfer that can allow better public oversight of government while allowing more efficient delivery of services. I will provide supporting evidence for this view in this chapter but, once again, strike a note of caution. For the Internet to be a powerful tool of government service delivery requires much more than the physical presence of a network of computers, as numerous examples of failed e-government projects make clear. But even the expense of only the first step of access is likely to limit use of the technology as a direct provider of services such as health and education in developing countries. Direct rollout programs aimed at bringing Internet access to poor, rural communities are particularly prone to limited returns—the nature of poverty in the developing world makes the Internet an ill-suited tool of direct poverty reduction.

Government Interactions with Business and Citizens Through the Internet

There are numerous examples of ICT-enabled government efficiency improvements from every corner of the world. In Singapore, where the government spends approximately US$100 million per year on ICTs for the civil service, studies suggest that every dollar spent on this program has generated US$2.70 in returns due to expanded productivity and reduced operational costs.[1] One dramatic example is the country's Tradenet networked information system, which allows importers and exporters to declare goods from their office computers. The information is routed to various government departments, and clearance is usually received within fifteen minutes. Combined with a real-time vessel management system at the port, it has allowed for ship turnaround times of less than ten hours and estimated savings of about 1 percent of the city-state's GDP. Every year, the average customs official in Singapore is able to clear $666 million worth of goods. Another electronic trade documentation system based on the Tradenet model is GCNet in Ghana, which has reduced average clearance times at KIA airport from three days to four hours.[2] Compare this performance with that of other developing countries—in Egypt, imports still face three days of delays going through clearance procedures, and the average customs official clears just $600,000 worth of goods a year. These delays cost the equivalent of a 15 percent tariff on imported goods.

Networked computers have also played major roles in improving the efficiency of government purchasing. When Mexico implemented the first phase of its online procurement system, for example, it is estimated that average savings of at least 20 percent were realized due to the increased competition and transparency of the system. In Chile, savings from e-procurement were estimated to total $200 million, or 1.4 percent of government expenditure, each year.[3]

A number of developing countries, including Chile, have seen widespread use of online tax payments systems. The Internet can also simplify government-business relations. Mauritius's Contributions Network Project allows several hundred firms to pay income and value-added taxes over electronic data interchange (EDI) and web-based systems, as well as meet tax reporting requirements through one portal.[4] In the Zhongguancun Science Park in Beijing, 6,000 firms use a portal that consolidates thirty-five government procedures spread over twelve government departments, including licensing, financial reporting, and tax statements, reducing turnaround time by two-thirds.[5]

There are even cases of poorer developing countries such as India providing Internet-based applications that reach down to the individual citizen. Karnataka (a state in southwestern India) introduced an electronic land-title system called Bhoomi in the late 1990s, which delivers titles through kiosks installed in land offices. Obtaining a title once required up to thirty days, and usually a significant bribe on top of that. Today the title can be obtained for a fixed fee of thirty-two cents in less than an hour. The 800,000 people who use the Bhoomi system each month benefit from a process made far more transparent and efficient through the use of a well-designed restructuring involving the heavy use of ICTs. Consumer surplus and producer savings from reduced bribes and increased efficiency are estimated to equal approximately ninety-four cents per user.

This successful program has several noteworthy features. First, it involves computerization of an existing system that was in high demand and of great importance to the population (land records were required to obtain loans and insurance). It is not a case of constructing a network with the hope that demand would follow. Second, the major benefit from the project is reduced reported corruption because of increased transparency, rather than increased efficiency per se. The economic return on the project, excluding consumer surplus from reduced costs of bribery, is closer to twenty-three cents per user—still significant, but less dramatic. Third, at the same time as computerization, service delivery points were reduced from 9,000 to 177 across the state. After removing reduced wage payments caused by the falling number of service delivery points, the economic return of the project that might be more accurately seen as directly related to efficiency gains from computerization is eight cents per user. Furthermore, the reduction in service delivery points has led to complaints from some users of considerable additional travel costs. In short, the benefits from the program have to do with increased efficiency due to computerization, combined with a thorough reform of business processes, staffing, and infrastructure related to land titling. The major savings resulted from a reduction in service outlets and a reduction in corruption due to new processes introduced alongside and with the aid of computerization. Finally, it is worth noting that the project to computerize land records took from 1991 to 2002 to complete, suggesting the immense complexity of such an operation.[6]

This example yet again shows that the success of an Internet application depends on the wider environment and broader processes of change. Robert Wade of the London School of Economics, criticizing an article (that I coauthored) on the impact of the Internet in developing countries, notes that those selling the power of the Internet as a development tool often take the fact that an application has the potential to achieve something as evidence that it will or has. In this case, the application involved a networked computer system in Andhra Pradesh, a state in southeastern India, which had the potential to reduce the time needed to obtain a caste certificate (essential for obtaining government services) from as many as thirty days to ten minutes. Wade's "contacts in the capital city of Hyderabad, queried about this statement, politely suggested that 10 minutes might be the technological minimum, but not the real time facing any real applicant except possibly the minister's son. Not 10 minutes, and not 10 days." Again, the point is not that using networked computers had no impact on efficiency—it almost certainly did. The point is that the impact was limited not by technical constraints but by the usual institutional, policy, and other factors. Use of networked computers does not happen in a vacuum.[7]

Following on from this evidence, the conclusion of a recent survey of e-government Internet sites designed to provide information to citizens in developing countries should come as no surprise: "Introducing e-government without the corresponding institutional reform of the civil service system and organizational reform of the agencies may only lead to limited success in enhancing accountability," note Wilson Wong and Erich Welch, after analyzing 267 websites from governments across the world.[8]

The examples of failure caused by an inhospitable broader environment are legion. To take specific examples, in India, a program established by the National Informatics Center to provide ICT rollout and support to local governments for the storage of land records and monitoring of Ministry of Agriculture programs found that after fifteen years, the program had had only limited effects. As one study notes, "The impact . . . on administration has been marginal because the task of changing the administrative culture is enormous. Although IT can be a tool for decentralized planning, integration across departments and reduction in workload, it cannot be the sole instrument of change."[9] Similarly, the Ministry of Income and Taxes sought to computerize a major component of record keeping and storage. Yet political antagonisms between regional tax commissioners and the central tax board meant that most of the project never became operational, and only very limited achievement of reform objectives occurred.[10] In South Africa, the introduction of an intranet system providing information on property to staff and clients of the Johannesburg Metropolitan Council remained unused because its introduction had been technology-focused, rather than needs-focused, and training was inadequate.[11]

In general, Richard Heeks's work, which suggests 80 percent failure rates for government IT projects in LDCs, indicates that such projects are even more prone to problems and failure than they are in the developed world. Poor technology choice, poor institutional design, weak capacity to operate—all can dog projects. More generally, human, physical, social, and institutional capital-intensive projects, such as networking government operations, are likely to be ill-suited to environments lacking all those types of capital.

E-government applications that attempt to reach populations at large are particularly expensive. Total hardware costs for a one-computer Internet office with wi-fi connectivity have recently been estimated at $1,250 on-grid, rising to $5,000 off-grid, and over $7,500 if the only option is VSAT connectivity.[12] These capital costs are particularly high in relative terms. We have seen that the average cost of ICT in Africa is higher than in the developing world, but the average public sector wage is as little as one-tenth of its Western equivalents.[13]

Take the example of an e-health application in The Gambia. Nurses on a remote river island use a digital camera and a laptop to photograph visible symptoms that they cannot recognize or treat. The images are transferred to a physician in Banjul, who either prescribes a treatment or forwards the images via e-mail to a company in the UK, which can access specialists around the world and report back findings.[14] That is a wonderful story, but given an average annual health expenditure in The Gambia of $46 per head, it can hardly be rolled out countrywide.

Improvements in health are closely connected with education—women's education, in particular. This education might provide people with the information they need to make better choices involving child spacing, or washing hands, or using simple medicines like oral rehydration packs (basically a package of sugar and salt). Cheap technologies that require little in the way of supporting infrastructure, combined with the education allowing residents to access such technologies, may improve the quality of life. But we have seen the Internet is not a cheap technology that requires little in the way of supporting infrastructure—once again, then, the Internet appears to be an unlikely technology to foster radical change in health outcomes in developing countries.

Indeed, intermediary uses of the Internet may benefit the poor most. In a group of sub-Saharan countries referred to as the "meningitis belt," the World Health Organization (WHO) has implemented an electronic system by which daily reports of disease outbreaks are relayed to health professionals, who collate the data and use findings to target mass vaccination programs. Personal digital assistants have found a role as data input devices for volunteers carrying out health-related surveys in Ghana. Such uses are likely to be more sustainable and may make a major difference. But given that many countries lack even the resources to provide vaccines to their populations, it is unlikely that the Internet alone will spark a medical revolution.

Overall, given high costs and the high risk of failure, the rollout of e-government applications, especially to the level of the individual citizen, may not carry as high returns as a few well-advertised but underanalyzed success stories such as the Gyandoot example from Chapter 1 might suggest. There will be roles for networked computers even in poor countries—both in "back office" functions such as budget management and in relations with more established businesses. Perhaps one of the most obvious of these involves promoting transparency, where putting information up on the Web allows some to access that information and pass it on through other media that are more widespread. Nonetheless, direct to citizen e-government is unlikely to be a practicable solution to the problems of governance and the provision of government services in poorer developing countries in the near future.

Education Online

Given the excitement that surrounds the use of the Internet as a tool for education in developing countries, it is perhaps worth looking particularly closely at its performance in that area. First, it is important to note that the effectiveness of well-designed, traditional, distance education programs using interactive technologies has been demonstrated in many different environments.[15] A recent survey of seven studies looking at the cost-effectiveness of different educational interventions suggests that, in terms of incremental improvement, the impact of a dollar spent on interactive radio instruction (in which broadcasts are combined with class participatory exercises) is nearly 70 percent higher than a dollar spent on purchasing textbooks and over eleven times higher than a dollar spent on teacher training.[16] Studies of Mexico's television- and radio-based *Telesecundaria* program have found that the program is only 16 percent more expensive per pupil served than normal urban secondary schools, and students benefit from much smaller student-teacher ratios and catch up significantly with their urban counterparts.[17] To date, however, this type of positive evidence is more scarce regarding the Internet.

The evidence of the impact of the Internet on education is, again, concentrated in OECD countries. In those countries, results are more mixed than usually assumed or proclaimed by proponents. As early as 1990, former US vice president Al Gore was using Internet imagery that would be cliché by the middle of the decade and far predates talk of a new economic order based on e-commerce: "a school child could plug into the Library of Congress every afternoon and explore a universe of knowledge. . . . A doctor in Carthage, Tennessee, could consult with doctors in the Mayo Clinic in Minnesota on a patient's CAT scan."[18] Yet the actual impact of putting networked computers in classrooms has been muted. Much as the prediction that TV would revolutionize the classroom now looks somewhat naive, so soon will claims that the Internet will do the same thing.

A recent study of the impact of the E-Rate subsidy program in California, which provided significant incentives for schools to wire their classrooms from 1996 to 2000, estimated that there were about 66 percent more Internet classrooms than there would have been without the program. The study concludes, however, "The increase in Internet connections has had no measurable impact on any measure of student achievement."[19] It is but one of a number of studies that point in the same direction. As Steve Jobs of Apple Computer (a company with a strong commitment to computers in education) has concluded, "What's wrong with education cannot be fixed with technology, no amount of technology will make a dent."[20] Again, I do not mean to say that the Internet will have no impact on education—especially distance education. Already, thousands are getting degrees online. But the impact will be marginal rather than revolutionary, mixed rather than purely positive. That is a sentiment increasingly echoed by users. Although parents seem happy with the fact that their children are spending time online, those actually trying to use the Internet for learning are becoming increasingly disenchanted. In 2000, 50 percent of those surveyed in a Pew Internet and American Life poll thought that the Internet helped "a lot" in improving their ability to learn new things.[21] By 2001, this percentage had dropped to 31 percent. Among those with the longest experience online, the figures were even lower.

Turning to the rest of the world, just as with the use of the Internet in business, the costs of putting the Internet in schools are likely to be higher and the benefits lower than in the developing world. In an environment where human resources for education are relatively cheap (teachers in LDCs are paid very little compared to their counterparts in developed countries) and IT capital is more expensive (computers tend to cost more in LDCs than in the United States), the cost-effectiveness of placing IT in schools (as compared to decreasing class sizes or using funds for interactive radio instruction) is likely to be lower than in the United States. One has to wonder about the relative net benefits of a project to bring the Internet to schools in Africa, when a recent education project in Uganda was hailed as a great success because it reduced the student-to-book ratio from forty to six students per book.[22]

Little evidence exists on the impact of the Internet on education in LDCs. Perhaps the most famous educational experiment involving the Internet in developing countries is the Hole in the Wall project. Dr. Sugata Mitra, of India's NIIT computer training and software services company, put a computer in the wall of a slum area of Delhi with Internet access, turned it on, and left it running. Within days, children were playing games on Disney.com and drawing with Microsoft Paint. The project was hailed by CNN and the BBC, as well as print media from the *Guardian* to *Business Week*, as showing that, without formal instruction, kids could master the Internet—and in turn that this computer literacy could spark a social revolution. The project won the UK Institute for Social Inventions 2000 Award for Social Innovations, and Mitra talked of taking the project nationwide, at a cost of US$2 billion.

It is worth noting, however, that the children spent the great majority of their time playing with games and paint programs. And unlike the Institute for Social Inventions, the parents of the children exposed to the experiment had distinctly mixed feeling about it. As one complained: "My son used to be doing very well in school, he used to concentrate on his homework, but now he spends all of his free time playing computer games at the kiosk and his schoolwork is suffering."[23] In other words, unsupervised use of the Internet by kids in India produced pretty much the same results as unsupervised use of the Internet by kids in the United States.[24] Why that should surprise anyone or lead to dramatic social improvement is not necessarily clear.

Regarding the Internet in schools, any of the more substantial pilot projects involving the Internet in education do typically incorporate some form of review and feedback. However, frequently they are limited to questionnaires asking project beneficiaries to describe their experience with the new technologies. For example, in a report on a recent major initiative to bring the Internet to LDC classrooms, the authors note that their measurements are based on subjective interpretations of the impact of the program, derived from questionnaires distributed directly to teachers, students, and administrators.[25] Although obtaining first-hand feedback is critical to ensuring that projects meet local needs, objective criteria are also needed to measure the degree to which perceptions adequately capture real improvements in learning.

One of the very few studies of the cost-benefit ratio of using computers in developing country educational systems is based on a randomized trial of remedial education in the slums of Mumbai and Vadodara in India, but it is not reassuring. It found that hiring extra teachers to provide remedial instruction was seven times more cost-effective in improving test scores than a computer-assisted learning program utilizing educational games.[26]

With regard to the impact of learning computer skills on future earnings potential, there is evidence from LDCs, as well as from the developed world, that computer users receive a wage premium over noncomputer users.[27] That might imply that market measures suggest the benefit of computer-aided education. Again, however, the evidence on the wage premium for computer users is mixed—many studies conclude that more highly paid workers use computers, rather than that computer use leads to higher pay. As we have seen in the developed world, even though computer usage and the sophistication of that usage correlate with increased income, measures of computer skills do not.[28] Further, even if a limited number of computer users in developing countries earn premium wages, that premium is declining over time.[29] Expanding computer access might speed that decline to the point that it is possible that supply of computer-aware workers will outstrip demand.

Arguments over the effectiveness or otherwise of the Internet on education in developing countries are largely moot, however. There just is not the cash to put computers in every school worldwide. In some low-income countries, the "discretionary" budget (what is left over after paying teachers' salaries, needed for such items as chalk and buildings) is about $5 per year per primary student. Primary teaching materials expenditures range from as low as three cents per student per year in Swaziland to $1.60 in Mauritania and $3.60 in Suriname.[30] Funding for secondary schools is slightly better, with discretionary expenditures perhaps two to three times larger. But that still suggests average expenditures of perhaps $20–30 per secondary student per year in low-income countries and $100–150 in lower-middle-income countries.

The cost of providing Internet access is far higher than can be supported by these budgets. Looking at total cost estimates from LDCs, three computer-assisted instruction projects that put computer laboratories in schools at the level of one computer per twenty students carried an annual cost of $78–104 per student.[31] It should be noted that the schools connected in these studies were located in wealthier LDCs—they did not face the problems of lack of electricity or phone service and faced less significant bottlenecks in terms of skilled technicians and physical access to schools. Further, these costs did not include Internet access itself.

Nonetheless, compare these figures to the budget available for funding Internet classrooms. Computer labs cost between 2.4 and 21 times the discretionary budget per primary student. If all of Zimbabwe's discretionary spending at the secondary level was used to provide students with computer access, it would buy perhaps 113 hours a year per student in front of the computer. The remaining (approximately) 1,240 hours per year would go unfunded, and there would be no money left over to buy books, chalk, or even buildings to support lessons for those thousand-plus hours.

The Internet may have a significant role to play in improving the quality of tertiary and vocational studies in developing countries, perhaps on the model of the African Virtual University—which operates learning centers in thirty-five universities across the region, providing online and video content to supplement face-to-face teaching. At the same time, it is unlikely the Internet will spur a revolution or will see widespread day-to-day application in education in developing countries.

Public Access for the Poorest

We have seen that intermediary uses of the Internet have, in a few cases, proved themselves effective in enhancing business opportunities or the delivery of government services to relatively poor people, even in low-income countries. There is a considerable further step from intermediary uses to direct use by the great majority of the poor themselves. Nonetheless, governments worldwide are dedicating themselves to the goal of universal access—even to the poorest of the rural poor—spurred on by the international community.

Indeed, numerous reports highlight the digital divide in terms of individual usage as a significant development problem.[32] The assumption appears to be that lack of access to the Internet is a significant bottleneck to development for poorer people and across poorer countries of the world. Higher usage per capita would be a good thing, and ubiquitous usage would be a great thing. This assumption, made in reports on the digital divide in the United States—reports such as *Falling Through the Net* by the Department of Commerce—has been imported wholesale into discussions on the global digital divide. And it is shared by development practitioners, industry leaders, and policymakers—the G8 at Okinawa, Silicon Valley CEOs at Seattle, and politicians at the UN.

There are considerable digital divides within countries that do suggest far lower access than is available in developed countries, especially at the household level. In particular, getting access to advanced services in rural areas presents a challenge. Although the mobile footprint covers over one-half of the rural populations of the developing world and access to public telephony has spread incredibly rapidly, even in rural settings, only 10 percent of Thailand's Internet users are rural.[33]

But in part because coverage is low and rural areas have low population density, ubiquitous usage would require a lot of money. In this section I show that the costs of providing global "universal access" to the Internet would be very large, especially compared to the resources that are available to poor people to improve their lives. And when poor people have the option, it appears that they would rather use those scarce resources for things other than the Internet

Maybe poor people are demonstrating their "knowledge gap" here: if only they knew how the Internet could transform their lives, the poor with access would all be using it. Or it may be instead that those calling for universal Internet access do not fully understand the challenges poor people face and the environments in which they live.

In 2001, a cooperative venture between MIT, Microsoft, Alcatel, and the Costa Rican government set out to show that a rural telecenter project could make a difference to the quality of life of the rural poor. Little intelligent communities (or LINCOS for short) would provide the poor with access to telemedicine, distance education, and a range of other services. Some of the original participants wrote up their success, noting that they had developed a rural center that did not require access to telephone lines or the electricity grid (utilizing solar and satellite technologies) at an investment cost of only $20,000 per computer (it should be noted that this sum is approximately equal to the total annual incomes of fifty-five people living on a dollar a day).[34]

Two years later, the project is abandoning its rural base for the town of San Marcos because the intended beneficiaries have stayed away. Not that the project was an all-round failure—local coffee farmers, although by no means the poorest people near the village of El Rodeo where the center is based, have used the center to check coffee prices and register trademarks.[35] Once more, it appears that the Internet is far from irrelevant to developing countries, but not everyone sees the need—or perhaps has the need—to access it.

We have a flood of evidence suggesting that improved information flow can help poor people and that they recognize that fact and are quite willing to pay for that information. For example, the poor in Chile consider telecommunications such a basic service that they spend more of their income on telecommunications than on water. Furthermore, the *average* Chilean spends more of his or her income on telecommunications than on electricity and water combined.[36] Worldwide, even the poorest people appear willing to spend 2 percent or more of their income on telephony, if it is available. People with access to telephones have an easier time setting up small enterprises, pay less for crops, receive more for their own agricultural produce, and have better access to services—as supported by any number of surveys and studies since the 1960s.[37] With advances in mobile telephone technology—not least text-messaging—poor people can use their phones to access a growing number of services, including mobile banking. Poor people also access information from broadcast and print media. Greater access to such sources of information has been correlated with improved development outcomes in numerous surveys.[38]

Networked IT solutions are also being used in ways that directly affect the poor. In South Africa, an automated savings system aimed at poor depositors called AutoBank E is attracting thousands of customers who previously could not or did not have a bank account. An opening deposit of only US$8 provides customers with a wide range of electronic banking services through ATMs. In addition, the bank has used the data collected on depositors to analyze credit-worthiness, improving credit access for the country's poorest citizens. The system is very popular, with 2.6 million depositors in 2001.[39] It is but one example among many of advances in communications and information technology improving the lives of the poorest. But it is worth noting that it is not the same as providing full Internet access to every citizen. Instead, it involves providing a limited range of financial services to a wider range of people through an existing network. That might be considered an appropriate technology, in contrast to full Internet access, which is perhaps less suited to the needs and capacities of the poor.

The issue of appropriate technology is well illustrated by efforts to get price information to farmers. In the Gobi region of Bayanhongor province, 600 kilometers from the Mongolian capital of Ulan Bator, Mr. Daavadorj makes his living by herding livestock. In October 1999, he started listening to a ten-minute farm radio program. The program broadcast the prices of cashmere, skins, and petrol, along with twenty-seven other commodities in local markets, on a twice-weekly basis. "When we had no such program, we had to sell livestock at the prices traders offered," reports Mr. Daavadorj. "But now that we know the real price [in the markets], we just hold back until a fair price is offered."[40]

About 3,000 kilometers to the south, Halima Katuun is an illiterate woman who sells eggs for a living in rural Bangladesh. She recounts the impact that access to a mobile telephone has had on her business.

> I always sell eggs to middlemen. In the past, whatever prices they offered, I accepted because I had no idea about the growing price of eggs . . . Last week, the middleman came . . . and desired to pay me 12 taka per hali [four eggs]. . . . Keeping him waiting, I rushed to check the prices through the village phone. The price was 14 taka for four eggs in nearby markets. I came back and refused to sell to him at the lower prices. . . . After a brief haggling, we agreed to buy and sell at 13 taka per hali.[41]

Across the Pacific Ocean in Chile, the national agricultural extension service has created an Internet-based rural information service for farmers' groups, rural governments, and NGOs. As the Food and Agriculture Organization (FAO) reports: "It was estimated that transmitting price and market information this way cost 40 percent less than using traditional methods. In addition, the information was more timely, reaching farmers much faster. In the past, the publication and distribution of a printed bulletin took 45 days."[42]

All three of these stories regarding price information show the importance of information to the rural poor. Knowing the market price of produce puts farmers in a far stronger position vis-à-vis middlemen, allowing them to earn more for their output. But these stories also show that information can be delivered in a variety of ways. In this case, radio, telephone, and Internet are all being used to convey essentially the same data to farmers. And although the FAO argues that the Internet has proven to be faster and more cost-effective than previous techniques to disseminate price data in Chile, that does not mean that it is the *most* effective method.

The case from Mongolia is enlightening here. In addition to broadcasting data on prices available on about 400,000 radios countrywide, Gobi Business News also posts the data to its website. Since the project website was launched, the price data has been accessed an average of just fifteen times for each biweekly update. It is pretty clear which method of data transmission is more likely to reach the rural farmer. Radios—with wider access, lower cost, and less demand for technical skills and literacy—are probably the most cost effective method for delivering such information to the rural poor. Telephony might run second. It is probable that the Internet is a distant third. By and large, poor people lack access to the necessary infrastructure and equipment for Internet access, and providing that access would be expensive. At least as important, they lack the skills and the enabling environment required to reap the benefits of access.

Turning first to access to infrastructure and equipment—a telephone line and electricity—the majority of the poor do not have home access. In South Africa in 2000, 0.6 percent of households in the poorest quintile owned a fixed telephone (as compared to 75 percent of the richest income quintile).[43] In Tanzania, only 0.5 percent of the poorest households have electricity.[44]

One reason for the current low rate of access to networked infrastructure such as electricity and telephony, and also a reason to assume that it will be very expensive to provide in the future, is that 69 percent of the population in low-income countries is rural.[45] It is far more complex to provide networked services where population densities are low (and even more where the geography is difficult—mountainous or forested). Fewer consumers per square mile translate into the need for much more infrastructure per person and higher costs. Combined with lower incomes in rural areas, these factors make the economics of rural network provision unfavorable.

Universal access programs that focus on the areas that are least economical to serve provide an indication of the greater complexity of providing network access to rural and remote areas. Perhaps the most efficient recent attempt to provide universal telephone access to citizens of a country occurred over the last few years in Chile. There, the government auctioned subsidies to the lowest bidder among private companies that committed to provide public telephone services to areas previously lacking access. The average subsidy countrywide required to provide public access was about $10 per newly served person. But subsidies rose to around $100 per person in areas with a population density of under ten people per square kilometer.[46]

Such population densities are common in countries that are home to the world's poorest people. A brief look at the atlas suggests that there are at least nine African countries where the majority of the land holds fewer than *one* person per square kilometer.[47] The majority of land in most of the rest of the continent is occupied by fewer than ten people per square kilometer. The same is true of Latin America and Asia—with the exception of the Indian subcontinent and the eastern provinces of China.

The Internet requires more than just telephony, which we have seen is spreading rapidly, even in rural areas, and is perhaps the least significant barrier to expansion. At the least, the Internet requires an immediate source of electricity, computers, an ISP connection, and skilled technical support. Thus the economics of Internet provision in rural areas are considerably more complex than those for basic telephone service. As we have seen, the LINCOS off-grid solar-powered telecenters have a setup cost of approximately $20,000 per computer. Ongoing costs, excluding manpower, were approximately $5,000 per computer.[48] If capital costs are spread over five years (with no allowance for interest payments), the annual cost would be $9,000 a year. Assuming, based on figures from Ghana, that a technician to support a six-computer center would command a salary of at least $6,000, that brings annual costs per Internet-enabled computer up to $10,000.[49]

Complexity does not equal impossibility. All sorts of other creative ways to get the Internet to far-flung places are being tried in LDCs. In Brazil, they have put the Internet on the river. The Navegar project has eight desktop computers, a GPS system, a digital camera, a scanner, an inkjet printer, two Web cams and a satellite dish inside a three-floor, wooden boat. The boat travels around the eight-island archipelago of Bailique, serving about 8,500 people in thirty-eight communities.[50] In Malaysia, the Internet is available on a 40-foot bus, which has been used to provide computer courses to 2,800 children over two years in rural areas around the city of Tunjang. Before taking this computing course, only about 9 percent of students were able to use e-mail; afterward, just over 50 percent of students could use it. But, again, these solutions come at a high cost. The bus, for example, is valued at $263,000. Capital costs per student successfully taught how to use e-mail run at perhaps $230.[51]

Compare this figure to potential communications expenditures among poor people in developing countries. The poorest quintile in Chile are willing to spend approximately 2–3 percent of their income on communications.[52] Assuming the same holds true for the 1.5 billion people worldwide living on a dollar a day, it suggests maximum yearly communications expenditures of approximately ten dollars. If *all* this expenditure went to Internet access, then 1,000 users would be required to support each Internet-enabled computer in a setup like LINCOS. If the technician kept the center open eight hours a day for 300 days a year, each user would have one two-and-a-half-hour slot per year to access the Internet. At a population density of ten people per square kilometer, the average user would have to walk about two hours each way to reach the center. At a population density of one person per square kilometer, the walk each way would take most of a day. It is not clear that this level and quality of Internet access is either sustainable or that valuable.

LINCOS is known to be an expensive model, and rural connectivity in easier environments carries considerably lower costs. Earlier in this chapter, I mentioned that off-grid access provided by VSAT can cost closer to $7,500 per computer, a little over one-third of the cost of the LINCOS model. Furthermore, it is likely that the costs of rural Internet provision will fall dramatically over the course of the next few years. Solar power, computers, and communications networks connections are all dropping in price. Combined with economies of scale in the provision of telecenters, costs might drop by as much as two-thirds.

Nonetheless, the physical costs of a full networked computer setup are unlikely to drop too far below US$1,000 per access point, even in relatively population-dense rural areas with electricity access.[53] Such figures still suggests the average user would have access to the Internet for a total of perhaps eight hours per year. Although this might be enough for basic communication needs, it would not support extensive e-learning, e-commerce, or even entertainment uses of the new technology. Even assuming no other barriers to the technology's use, eight hours of Internet access a year would not be a "transformative" level of connectivity by any means. To support a transformative level of access for the poorest would require significant subsidy rates. One hour per week of access, for example, would require a subsidy of over $50 per capita per year for those living on a dollar a day—or about ten times public spending per capita on health in low-income countries.[54]

Problems with the cost and complexity of physical access to the technology of the Internet are probably not the most significant barriers to high utility of the technology for the poorest, however. For example, the average person living on a dollar a day is illiterate.[55] The Internet has the power to convey sound and video (although that power is much diminished in an environment of slow connections and frequent line drops), but much of its information is textual. Given that, illiteracy poses a major barrier to Internet use.

The majority of those living on a dollar a day also speak a minority language in their own country. Even fewer speak English, in 2001 the language of 72 percent of the world's websites.[56] Many of the languages of the poor are almost completely absent from the web. Take Igbo (Ibo), a language spoken by 17 million people in Nigeria. A search that I conducted in 2002, lasting two hours—or about the average yearly length of access currently affordable by people living on a dollar a day—came up with just five sites: a translation of the Universal Declaration of Human Rights; a translation of a religious document called "the four spiritual laws," a translation of the food pyramid, a two-page Igbo phrase book, and a prayer manual. By 2006, there were additional sites on learning the language but still under fifty sites in Igbo for Igbo speakers. There are no sites offering an automatic page translation service from English to Igbo, and so an Igbo speaker would be limited to these few pages on the web. It might be questionable if access to these five documents is worth $10,000 per enabled computer per year.

Not surprisingly, given the lower utility of the Internet to speakers of minority languages, there is an underrepresentation of non-English speakers among the world's Internet users, as we have seen. It is also unlikely that the disadvantaged status of poor, minority-language Internet users will change in the near term. There is a high fixed cost of website creation (as demonstrated by the number of failed website companies in the recent dot-com crash), and a small language group of predominantly illiterate people living on a dollar a day is hardly an attractive market for e-commerce or content providers. Again, it is unlikely that the necessary technical skills for Internet use will become widespread among the poor in LDCs in the near future. Discretionary budgets per student in LDC primary schools are as low as five dollars per year, as we have seen—hardly enough to support a computer center.[57]

Even if the human and physical capital barriers to Internet use can be overcome, it is doubtful that the utility of the Internet for the poor in LDCs will be as significant as it is for the wealthy of the world. First, the poor are less reliant on market transactions made (somewhat) more efficient by the Internet. In Tanzania, for example, 83.5 percent of the poorest are crop producers—much of that for subsistence. Fifty-six percent of this group's food expenditures, their largest budget item, are in kind rather than cash.[58] Even when poor people do make market transactions, they do not have access to the financial infrastructure (credit cards or bank accounts) to make purchases online, nor is the physical delivery infrastructure (roads, postal networks) in place to make direct business-to-consumer transactions a feasible option. These weaknesses account for the fact that only 2.2 percent of India's Internet subscribers have engaged in e-commerce activities.[59]

The barriers (physical, human, and institutional) to Internet use among the poorest in LDCs are significant, then, which might help to explain rural preference for communications technologies other than the Internet. Seventy-one percent of rural people surveyed in Nepal used the radio as a source for information and found it effective, for example. Friends, family, and political leaders were the only information sources ranked as more effective. Compare that to the telephone (used by 19 percent, but only found effective by 1 percent), and the computer (used and found effective by only 1 percent).[60]

The nonphysical barriers to the Internet also help to explain the low usage of Internet facilities, even in poor areas lucky enough to have access. In urban India and South Africa, perhaps one-quarter of the population has access to the Web, but only 10 percent are regular users.[61] In Indonesia, between 20 and 25 percent of the population have access, but only 5 percent are using the Internet.[62]

Among the rural poor, user rates are even lower. A recent survey of two villages in Uganda,[63] conducted at the time Internet-enabled telecenters were set up in the villages, suggests usage rates below 5 percent (compared to close to 30 and 100 percent for telephone and radio use, respectively).[64] Again, a survey involving villages in Gujarat (India), Mozambique, and Tanzania, all located near towns with Internet access, found very limited use, indeed. "It was hoped that this report would provide information about use of and attitudes towards the Internet," the researchers write. "In practice, however, in spite of the availability of Internet facilities in local towns, less than two percent of those surveyed had ever made any use of these." In contrast, about 90 percent had made use of a broadcast technology, and about 70 percent used a telephone at least five times a year.[65] A survey of a pilot program of the Ministry of Environment, Natural Resources, and Fisheries in Mexico found that of twenty-three telecenters set up in rural areas around the country, only five remained functional after two years. Problems encountered included insufficient maintenance funding and limited community interest in the projects.[66] In turn, low demand and high cost might help to account for the fact that in a recent survey, the United Nations Development Programme could find no examples of donor-funded telecenters that were fully sustainable.[67]

A perhaps rational calculation of the limited power of the Internet to improve their situation recently led a group of poor Peruvian farmers to rally against the construction of telecenters. The farmers, from the Apurimac region, were protesting the use of resources from a poverty alleviation fund run by the government to provide Internet access rather than improved water supply. One was quoted as saying that "many of us cannot read and write. . . . we don't believe the Internet will help us as much as an irrigation channel will."[68]

In the medium term, Internet access (and, in particular, utilization of that access) is likely to be concentrated among the wealthier (better-educated, creditworthy, transport-linked, and English-speaking) of global citizens, then. Given that the Internet can sometimes be a powerful tool for increasing incomes, it is likely that, if anything, the Internet will be a force for widening income inequality between the rich and poor.

Certainly, that appears to have happened with the telephone. We have seen that telecoms access really helps the poor. However, the potential benefits of access to the poor can only be realized if that access is present. Historically, that has not been the case, with an urban elite having the most access. The poor have limited resources to afford telephone services, and they are shut out completely in the absence of public call facilities. In a study I carried out with colleagues Emmanuel Forestier and Jeremy Grace, we found that telecommunications rollout was associated with increases in inequality over time. Telecommunications rollout was concentrated among the already wealthy, it was a power for increased income, and so logically (and empirically) it increased income disparities.[69] With the recent expansion of service coverage thanks to mobile rollout, hopefully this picture will be changed for the telephone. It is likely to be a good deal longer before it changes for the Internet.

Perhaps most seriously, from the perspective of equity, the use of the Internet in development programs will backfire if those who need services provided over the Internet are excluded by the technology. For example, Singapore and New York have begun providing government services directly to customers online. If this type of service provision becomes a substitute for rather than a complement to more traditional methods of receiving government support, those without access to the Internet will be excluded.

The answer here is not to set up rural Internet kiosks that allow cheap or free access to government services. We have seen that such kiosks will be very expensive to set up. Further, we have seen actual evidence that where such kiosks have been set up, they are sparsely used—as in the case of India's Gyandoot, discussed in Chapter 1. For wealthy, urban, literate, technology-savvy customers of government services, online access will be a boon. For poor, rural, illiterate citizens, it is likely that there are more efficient ways to improve the quality and delivery of government services than the Internet kiosk. Nonetheless, the rural consumer will be at a disadvantage, which is why the Internet may be a technology for divergence of incomes within as well as between countries, rather than a "leapfrog technology."

To conclude, government use of the Internet in developing countries will face similar hurdles as use in the private sector—limited physical and human capital and an institutional environment inhospitable to widespread adoption. The overall level of development will limit citizen access, but even more so it will limit use where there is access. That suggests potentially low returns on access programs and on direct government-to-citizen Internet applications. At the same time, it is important to note that all citizens, including poor people, benefit from improved flows of information. Although other technologies may be better suited for direct provision of such information to poor people, the Internet may have an important intermediary role. Where applications are introduced as part of a broader process of business process reengineering they may, in some cases, considerably increase the efficiency of "back office" government activities.

Notes

1. UNESCAP 1999.
2. Sudan 2005.
3. Schware and Dean 2003.
4. Heeks 2002.
5. Qiang et al. 2005.
6. Lobo and Balakrishnan 2002.
7. Wade 2001.
8. Wong and Welch 2004.
9. Bhatnagar 2000.

10. Lal 1999.
11. Africa Development Forum 1999.
12. Dot-Com Alliance 2005.
13. Heeks 2002.
14. ITU 1999b. See also Wright 1997.
15. Potashnik and Capper 1998.
16. Adkins 1999.
17. de Moura Castro, Wolff, and Garcia 1999.
18. Quoted in Cassidy 2002.
19. Goolsbee and Guryan 2002.
20. Quoted in Oppenheimer 1997.
21. Pew 2002.
22. Association for the Development of Education in Africa 1999.
23. Warschauer 2002.
24. Similarly, a survey of Internet users in an Internet café in Mamobi, Ghana, found that the overwhelmingly young users found that "*everyone* is chatting," either through instant messaging or chat rooms, using the Internet as a way to find electronic penpals—very few visit websites or use the Internet for research, and fewer still use it for business transactions. That may help to explain local educators' reactions to the Internet cafés as places where children should be banned (Slater and Kwami 2005).
25. Kozma et al. 1999.
26. Banerjee et al. 2004.
27. See Patrinos 2001 for a review.
28. Borghans and ter Weel 2000.
29. See Patrinos 2001.
30. Adkins 1999.
31. Reported by Potashnik and Adkins 1996. With higher student-to-computer ratios, these numbers obviously fall. For example, Cawthera 2001 provides estimates from Turkey about classrooms where there are forty students per computer, which suggests costs of $32 per student per year.
32. See, for example, UNDP 2001.
33. Ibid.
34. Shakeel et al. 2001.
35. A. Amighetti and N. Reader, "Internet Project for Poor Attracts Rich," *Christian Science Monitor,* July 24, 2003.
36. De Melo 2000.
37. See Egleston, Jensen, and Zeckhauser 2001; and Forestier, Grace, and Kenny 2001.
38. World Bank 2001b.
39. "South African Banking: Profit in Poverty," *Economist*, March 25, 2000, p. 81.
40. Croft 2001.
41. World Bank 2001b.
42. Balit 1998.
43. World Bank 2000b.
44. World Bank 1996.
45. World Bank 2001a.
46. See Kenny 2002a.
47. John Bartholomew and Son 1989.
48. Shakeel et al. 2001.
49. Africa Internet Forum 1999. India provides an interesting example of costs here. As of 2004, perhaps 1 percent of India's 600,000 villages were connected to the

Internet via one type of telecenter pilot or another, with the largest projects being e-Choupal, with 3,500 centers targeted at farmers, and n-Logue, with 1,024 wireless local loop–connected kiosks run by entrepreneurs. By at least some measures, the majority of theses centers are sustainable (the e-Choupals between them transacted $100 million in business in their first year). But scaling up these pilots to the national level will increase costs while decreasing potential revenues sources. Estimates for the marginal investment costs of serving 100,000 villages rise from $100 million for the first 100,000 to $750 million for the fourth 100,000, as population densities fall and geography becomes more complex. Under these circumstances, even the most profitable pilot will founder as it is scaled up (see Schware 2004).

50. Paulo Rebelo, "Keeping Email Afloat in Brazil," http://www.wired.com/news/wireless/0,1382,45919,00.html. .

51. Wayne Arnold, "Malaysia's Internet Road Show," *New York Times*, August 23, 2001.

52. De Melo 2000.

53. Kenny 2002a.

54. Estimated from World Bank 2001a.

55. Kenny 2002a.

56. Ibid.

57. Grace and Kenny 2003.

58. World Bank 1996.

59. Miller 2001.

60. Pigato 2001.

61. M. Patore, "Why the Offline Are Offline," at cyberatlas.internet.com.

62. Halewood and Kenny 2005.

63. For the International Development Research Center (IDRC), see http://www.idrc.ca/telecentre/evaluation/nn/22_Buw.html.

64. Not everyone who has access in the West feels the constant need to go online, either. A UK survey in 1999 found that, of people who classified themselves as "Internet users," 26 percent hadn't accessed the Net once in the previous week, and a further 20 percent had only accessed it once or twice. A good number of people "came, surfed and went back to the beach," to quote Wyatt 1999.

65. Souter et al. 2005.

66. Robinson 2000.

67. UNDP Evaluation Office 2001.

68. "Internet? Give Us Irrigation, Peruvian Farmers Say," Reuters News, October 12, 2005.

69. Forestier, Grace, and Kenny 2001.

6 — Sustainable Policies for E-Development

Cautionary Tales

Forty years ago, the following story was not uncommon: the government of an African nation borrows money from a development agency to build a steel mill (or car factory or conference center) in the country. Business plans suggest that, exploiting cheap labor and abundant local resources, the mill is a surefire winner. After it is built, the income from wages at the mill makes local people far better off. The project is declared a great success—it has raised living standards and pushed the country toward prosperity.

Of course, we now know how this archetypal story all too often concludes. The steel produced is never globally competitive, and the plant only survives because of large subsidies. Economic rates of return on the project are considerably negative.[1]

As of yet, we don't know if government-backed microchip plants and software companies are the twenty-first-century equivalent of the African steel mill. There are some good stories about the ICT industry creating jobs and opportunities. As with steel mills, the East Asian record with the ICT industry is one of considerable success. In some cases, the chip plants are employing hundreds on good pay. By and large, however, many a chip plant and software producer has garnered significant subsidies or tax breaks. It seems quite likely that the record of government-backed chip plants and software firms will be more similar to the patchy record of steel mills than to the perhaps Panglossian visions of some ICT optimists.

We should again be careful of assuming that the Internet is unique. For example, there is a technology for which the divide in access is large and growing. Already, there is more than a seventyfold difference in access rates between US and Indian households. That gap is far larger than the income divide between the two countries. Worse, the divide is linked to productivity, suggesting this differential access will promote ever-widening divergence in income. The divide I am talking about is, of course, the air conditioner divide.

As it happens, the air conditioner divide in 2000 really was almost exactly the same size as the differential between India and the US in the more famous divide—the digital divide in Internet users per capita (as of today, the cold fact is that the air conditioner divide is larger). Further, the air conditioner has been linked by economists with improvements in productivity (a finding that will surprise no one who has tried to work in 95-degree Fahrenheit heat at 100 percent humidity), and the air conditioner divide is in some ways more intractable than the digital divide (many households can share a computer, it is not so easy for them to share an air conditioner).[2] So why isn't there a G8 Air Conditioner Task Force?

Optimism over the power of the Internet to transform the developing world and concern over the scale of the digital divide has, not surprisingly, created calls for significant policy responses. We saw in Chapter 1 that Thomas Friedman has used the Internet to argue that the new institutionalist agenda (governance, education, free trade) is more important than ever. George Gilder goes as far as to argue that the Internet will (if we let it) create a utopia, with the two most significant changes being an end to regulation and the welfare state. Along with partisans of the left, the right has suggested that the development of the Internet only increases the importance of what they always said was sensible in terms of policies.

But the spread of the Internet has also driven some to suggest a whole range of policies specific to information and communications technologies. Donors have promised billions of dollars to roll out access to the Internet; advisers have suggested subsidies, tax breaks, and tech parks for Internet-enabled firms and computer companies; and presidents and prime ministers have created "e-czars" and announced that the Internet will be at the center of their education policies for the twenty-first century.

These statements and policy proposals are worth reexamining in the light of what we have learned about the actual potential impact of the Internet. The ICT industry, even while creating significant employment and investment opportunities, does not appear to be an especially valuable form of enterprise to the economies of the developing world. Internet use has spread rapidly worldwide because the many entrepreneurs who see advantage in the technology have embraced it—but entrepreneurs who do not use it, or use it only to a limited degree, might not be driven by Luddism but instead by a rational calculation of the benefits the technology can bring. Government use can bring benefits, but the technology is unlikely to be transformative and unlikely to be successful at all, absent broader institutional change. The poor may benefit far more, and far more sustainably, from other uses of resources than Internet-enabled telecenters. And, overall, the economic impact of the

Internet may be muted—as much a force for further divergence as for "leapfrogging."

Most in development circles would probably agree that the types of economic intervention that governments should favor are those that sustainably increase the incomes (or, more broadly, the quality of life) of everyone, but have an especially large impact on the poor. As we have seen, the evidence to date is that initiatives supporting Internet use are likely to have a limited economic impact, but what impact is felt will be on the rich. Nonetheless, this broad picture is doubtless oversimplified, and there may in fact be significant roles for government at both the national and international level.

Indeed, perhaps the most important thing to say about policies toward the Internet in developing countries is that there is no one right position. For "developing" countries such as South Korea, a policy of universal Internet access combined with subsidies for training, significant local content, and the development of government-to-citizen Internet applications is feasible and might well be advantageous. For developing countries such as Mali, the same program would be ruinous. If the record of past development silver bullets has suggested anything, it should be that one-size policies will not fit all. Nonetheless, some global conclusions may broadly apply.

Policies to Foster ICT Industries in LDCs

On the industry side, are there grounds for particular support of the Internet/IT industry over other industries in developing countries? A number of developing countries have taken a considerable gamble on this being the right decision: South Korea, Taiwan and Singapore all enacted national policies to attract the IT industry as early as the 1980s, with support for training, R&D, technology transfer, promotion, tax breaks, and IT parks, for example.[3] Malaysia launched perhaps the most ambitious plan with its multimedia supercorridor (MSC). The MSC began in the mid-1990s with promises of $10 billion worth of public infrastructure investment, a government venture capital fund, and significant tax breaks, among other incentives to ICT companies that would set up in the corridor.

From an economic perspective, what are the potential justifications for favoring ICT industries over the car industry or the steel industry, for example? Perhaps they are twofold. One is the type of leapfrogging argument that garners empirical support from impressive TFP statistics in the ICT manufacturing sector. As we've seen, however, this argument faces the problem that if there is a benefit from high TFP growth, it largely accrues to OECD consumers, not LDC producers. Whether it should be the first priority of the Malaysian premier to foster policies to benefit the US consumer is, surely, a question that might spark debate in Kuala Lumpur.

The second argument in favor of ICT industry support involves "cluster economies." That idea, somewhat worryingly, is a scaled-up economic version of the "first mover advantage" argument that pushed Internet firms such as Kozmo.com to expand rapidly before the business model was proved, which resulted in bankruptcy. Broadly, the cluster economy argument suggests that there is a strong advantage to ICT firms that position themselves geographically close to other ICT firms (perhaps because they can share a specialized workforce or benefit from ideas). So, countries that attract a few Internet firms may well see more firms coming to join them, which suggests a potentially high return on the small initial outlay of attracting a few firms through subsidy.

Evidence is mounting that advanced production of ICT goods and services does benefit from cluster economies—thus the emergence of areas such as Silicon Valley and Bangalore as centers of IT innovation. Looking at the United States in particular, industry data again suggest that IT-intensive industries do "cluster," in that the rate of convergence across regions in terms of employment in such industries occurs at one-half of the rate for all industries (in other words, areas that had more ICT employees in 1990 still have more ICT employees today). And the development of such a technology cluster can generate significant income—India now exports US$5.7 billion in software products per year.[4]

There may be something in the theory that ICT industries tend to congregate in areas already blessed (a reason why the Internet may be a force for divergence). However, it does not follow that government subsidy to ICT firms is a way to attract considerable industry at little cost. Clustering may have less to do with the presence of other ICT firms and more to do with the fact that all ICT firms are attracted to similar locations. The US study that found evidence of clustering, for example, suggested that it was not due to technology intensiveness per se. Instead, IT-intensive industries tend to rely on (unequally distributed) highly skilled labor, and those that do not rely on high-skilled labor see faster convergence—suggesting smaller clustering effects.[5] In the absence of a high-skilled labor force, then, there are few grounds for presupposing a "first mover" advantage—and so a role for government support for cluster development—in IT-intensive sectors.

IT clusters also tend to share a whole range of other features beyond a concentration of skilled workers, such as sources of venture capital, research centers, universities that foster practical research, and an economic and political climate that allows for innovation and attracts innovative people. Short-term government policies covering subsidies to ICT firms are only a very small part of the picture looked at by companies choosing to invest in a new venture.

Therefore, fostering clusters is likely to have an impact only where they are already forming in environments already suited to such industries. Indeed, such an approach does appear to have worked in parts of East Asia, where even critics of intervention admit the role that government support had in the development of the IT industry in Taiwan, for example (through public-sector laboratories, technology-transfer agreements, and even the creation of companies).

But even in East Asia, success has been mixed—Korea's attempt to expand computer production floundered because it supported the development of manufacturing for large systems just as the global market was moving toward PCs. Hong Kong's $100 million public venture capital fund for IT was forced to return funds to the treasury for lack of suitable investments.[6] The city's Cyberport initiative, which involved significant government incentives (not least a generous government land sale to the operating company), attracted only fifteen tenants in the two years after it was announced. A competing IT park in Hong Kong, opened without government support, is already six times larger than the Cyberport hopes to become.[7]

Malaysia's multimedia supercorridor is a particularly powerful case study here. The $10-billion-plus investment by the government was matched by just $475 million of private investment and 7,300 jobs (that works out at more than $1 million per job) up until 2000. Reasons that companies cited for not moving to the corridor included concerns about government monitoring of Internet traffic, capital controls, red tape, slow visa approval, weak intellectual property rights (IPR), and the absence of an appropriate skills base.[8] If such strategies sometimes floundered in the more successful parts of East Asia, it has to be a particularly high-risk strategy for a Rwanda or a Laos to follow.

It would, of course, be an even more risky strategy for Laos to follow if Rwanda was already following it, because of the potential for a race to the bottom in terms of competing subsidies and tax breaks of the sort that have frequently destroyed the economic benefits of export-processing zones.[9] The "fallacy of composition" (that what works for one country will work as well if every country tries it) begins to bite hard if the whole developing world decides to set itself up as a haven for programmers working on Doom Six or manufacturers of multi-terabit disk drives—as Russia and South Africa, also attempting to build IT industries, may soon discover.[10] The demand for new computers or programs or ICT-enabled services is finite. If the whole world starts to produce them, their price will inevitably fall, and less efficient countries (where the broader institutional environment is not as favorable) will be priced out of the ICT market. In turn, they will either have to subsidize the IT industry to an ever greater degree or see all their earlier subsidies come to naught. Such a race to the bottom would be a costly mistake.

It appears that this race to the bottom has already commenced, in the call-center sector in particular. In rich countries, most call centers remain in provincial cities, and attracting them to rural areas requires significant incentives. In the highlands and islands of Scotland, incentives have included subsidies to telecommunications service providers (this alone at a cost of over $2,000 per job created) and construction of low-rent facilities in government-owned business parks.[11] Many successful call centers in developing countries also appear to have garnered similar support. Dakash India, which has 3,600 call-center employees in New Delhi and Mumbai, operates under a regime of tax breaks and financial support that includes exemption from income taxes.[12] Software technology parks in India were provided with infrastructure, core computer facilities, ready-to-use office space, duty- and license-free imports, and tax exemptions, all at a significant cost to the government.[13]

A further reason governments may not want to attempt cluster formation is that, even if successful, they are likely to exacerbate inequalities by creating opportunities for people in regions that are already comparatively well off. We have seen that clusters generally form where there are already concentrations of highly educated people, venture capital, and other factors of success—and current wealth. Meanwhile, the spillover effects of clusters on poor communities are unclear. As Nobel Prize–winning economist Amartya Sen notes of Bangalore's software export industry, "even 100 Bangalores would not solve India's poverty and deep-seated inequality. For this to happen many more people must participate in growth. This will be difficult to achieve across the barriers of illiteracy, ill health and inequalities in social and economic opportunities."[14]

Governments should play some role in the development of local ICT industry. Perhaps most importantly, they can help create an environment attractive to venture capital and entrepreneurs. They can ensure access to quality information infrastructure (a topic to which I will return). They educate their citizenry, which supports human capacity building in some of the underlying skills needed for the exploitation and development of ICTs. (It is likely, however, that the ability of LDC governments to provide direct ICT human capacity building will extend only to tertiary education and its own employees because the cost of building a computer lab for every class is far too expensive.) There are a number of government functions that can frequently be provided more effectively with the help of information and communications technology, and governments can stimulate a local industry to develop and support such applications purely by taking advantage of such opportunities. But, given little evidence of supra-normal returns accruing to developing country producers, there appears to be little ground for sector-specific subsidies, tax breaks, government investment funds, or "IT zones." In a few countries at particular times, such subsidies and tax breaks may provide the required marginal incentive to promote industry development in an economically efficient manner. In most countries at most times, they will not.

Policies to Foster Use in Developing Countries

On the use side, a number of authors have suggested that governments in developing countries support increased access to the Internet. And calls for support have led to action—Korea required new buildings to be wired for broadband at an estimated cost of $1.5 billion, for example (apparently the greatest use of this access is for online gambling).[15] Even in Afghanistan, a number of donors and NGOs are supporting the rollout of Internet-enabled centers in Kabul and beyond, telecenters in post offices, telecenters for training women, and telecenters for telemedicine. Computers per capita has even made it in to the Millennium Development Goals as something the international donor community should monitor, at the same time we're measuring the percentage of people living on a dollar a day and the percentage with clean water.

Regarding poor people without access, perhaps it is most important to note that poor people do require information, but we have seen that the Internet is only one tool, and perhaps not the best one, to get it to them. Other communications technologies already more widespread, with less demanding capital and educational requirements, will be more suitable for exploitation. A range of evidence links telephony with improved quality of life in developing countries, and that technology faces far lower barriers in terms of education and language. Although we have also seen that providing access to telephony alone is expensive in sparsely populated rural areas, it is far more straightforward than providing access to the Internet as well and is likely to provide a similar level of benefit at much lower cost.

It may appear anachronistic to discuss the relative merits of telephony over the Internet in a time of convergence in communications technologies. It is true that telephone wires and fibers carry at least as much, if not more, data traffic than they do voice traffic, even in many developing countries. It is also true that most mobile phones manufactured in the past few years provide the capacity to input and transmit data (for example, text-messaging and camera phones). Looking forward, it is very likely that even cheap mobile phones will have ever-greater capacity. In turn, that means that functions such as mobile banking via text-messaging (already available in South Africa and being tried in Kenya) are likely to spread rapidly in the developing world.

But the fact of convergence does not alter broader realities. Advanced Internet applications that require stand-alone computers and broadband capacity are still likely to be stymied by weak institutions, low levels of education, and weak infrastructure. Convergence may make the mobile phone an ever-more-valuable tool of development, but there is significantly less evidence that it will do the same for Internet-enabled rural telecenters in countries where the majority lives on a dollar a day. If anything, convergence provides one more reason to focus on access to the simpler technological communications system of the telephone.

Radio is another technology that has already achieved near-universal access in LDCs: about 40 percent of rural households in the low-income countries of sub-Saharan Africa and Asia already own a radio, suggesting most people are at least near a radio set. Radios cost perhaps $10, plus the cost of batteries. They do not require an electrical connection or physical network connection. Programming is cheap enough to be produced locally and in a range of languages. On the transmission side, a low-power system can cost as little as $1,000. Digital sound recordings can be made on equipment that costs $800 or less. In central Mali, a station supported by Oxfam is broadcasting information to 92,500 people a year at a cost of just forty cents per person. Because community radio fills a role as a "community telephone," it has a major source of income. Even in a country as poor as Liberia, local stations have reached profitability—people are willing to pay $1 to have an obituary read on the radio, for example. Because radio is cheap and sustainable, in Latin America most radio (as opposed to television or Internet content) is produced locally or nationally. In Peru alone, an estimated 180 radio stations offer programs in Quechua, a language spoken by only 10 million people in the whole Latin American region (and a language that is almost completely absent from the Internet).[16]

Related to convergence, however, the Internet does have a role in supporting radio for development, acting as a distribution network among independent broadcasters. The Panos Institute's Banque de Programmes On Line, located in Mali, has correspondents in twenty francophone African countries, and Latin America's Agencia Informativa Pulsar is a similar initiative for Spanish-language programming.[17] Both these projects provide radio content accessible by community stations worldwide for broadcast.

In all, then, subsidies focusing on meeting the information needs of those living on a dollar a day in developing countries may well be better met through support of local and community radio stations or first-line telephone access than on the Internet. And tax breaks on e-commerce or training for learning the use of the Internet will likely benefit the rich at the expense of the poor, who will not be able to access the required technologies. Given that the economic benefit of wider use is yet to be conclusively demonstrated in developing countries, it may be that such subsidies will create not a rising tide to raise all boats, but instead a tide that will swamp the dinghies to the benefit of the yachts.

And although there may be an economic justification for government support of telecommunications infrastructure access, based on very high first-line costs of providing services to a previously unserved area, the justification for (consumer) usage or (producer) recurrent cost subsidies is far lower. If marginal cost is a significant barrier to Internet use, then there must be a small consumer benefit to marginal use. (If a three-cents-per-minute phone call deters people from logging on for another minute, it suggests that the marginal value to them of that access is less than three cents.) "E-rate" subsidies of the type used in the United States for schools and libraries, but emulated more broadly in countries such as Senegal, help the (wealthy) Internet user but are likely to provide little economic benefit to the country as a whole.

Furthermore, the type of investment subsidy scheme that has worked best in rolling out telecommunications access (discussed later in this section) does not work as well when paying demand does not cover recurrent costs. Chile, the country that pioneered the approach with pay-telephone rollout, found that when they piloted the same model for telecenters, only nonprofit groups applied—reflecting the fact that even if investment costs were completely covered by the government, rural telecenters even in upper-middle-income Chile were unlikely to be sustainable.[18] Limited evidence of demand, limited evidence of impact—especially on the poorest—and limited evidence of sustainability, all suggest that the public or publicly supported telecenter model should be retired.

Some commentators argue that the networked nature of the Internet creates a "threshold effect"—some level of rollout of the technology above which there begins to be a dramatic, self-sustaining growth in the utility of the technology. Below that level of rollout, they argue, the Internet is caught in the kind of low-use, low-utility trap we have discussed. That has been used as an argument to support public rollout of the technology to reach that threshold. It is a weak case. First, there is little evidence in practice of these low-use, low-utility traps preventing the spread of the Internet, even if they do inhibit the spread of more advanced uses. For such uses, the solution is broader development, not telecenters or other subsidies for access. Second, the theory flounders because the Internet is indeed a network, one that crosses borders. Unless one wants to suggest that the whole world is caught in a low-utility trap, the argument is considerably weakened. Third, even if you believe that national network size is itself a vital determinant, it would be very surprising if a credible level of public support for access would make that much difference—allowing a country to "cross the threshold." Of course, we don't know where this threshold actually is (if there is one), but before governments spend money trying to cross it, they should find out if they are below it, but close enough that their access program will make the difference. For the few countries in that happy position, Internet access programs may make sense, at least in theory.

Turning to the digital divide between enterprises in LDCs and high-income countries, the case for the necessity of Internet use is somewhat more compelling. Enterprises involved in international trade in particular might well need to use the Internet if they are to match Internet-enabled competitors. But we have seen that for businesses in LDCs, the "corporate digital divide" is one of use, not access. The incredibly rapid adoption of ICTs by enterprises in the developing world suggests that it would be hard to make the case for vouchers or subsidies being required to persuade ignorant entrepreneurs of the potential benefits of the Internet.

Again, the results of business surveys cited in earlier chapters suggest that the most powerful incentive for improving Internet use among businesses would be to improve the broader economic environment—infrastructure, institutions, policies. The World Bank recently conducted surveys of firms in East Africa to discover what those firms saw as the major constraints to doing business—part of the global surveys discussed in earlier chapters. A list of seventeen factors included corruption, macroeconomic stability, transportation, taxes, licensing, and labor skills. The list also included telecommunications. In Kenya, 40 percent of local firms and 67 percent of foreign firms mentioned telecommunications as a major bottleneck, placing it third for international firms and twelfth for local firms.[19] In Tanzania and Uganda (as in most of the rest of the world) telecommunications came last, mentioned as a constraint by under 6 percent of firms in Uganda, for example. The difference, especially regarding international firms, suggests two things about telecommunications policy in developing countries.

First, reform works. Tanzania and Uganda both have significant private involvement in the incumbent fixed-line telephone company. Uganda has some competition in the international segment. Kenya has neither. Second, reform works so well, it makes telecommunications a bottleneck for only a very small percentage of firms. Neither Tanzania nor Uganda has a nationwide backbone or universal broadband availability, but businesses do not appear to see that as a major constraint on their operations, perhaps because simple e-mail appears to be the "killer application" on the Internet in developing country settings. Before we spend public resources on extending backbones or broadband access because we see them as vital to sustain the competitiveness of African firms, then, it might be worth finding out if the firms themselves would rather the effort went to fixing potholes, improving skills, or redesigning the tax regime. Of course, it might also be worth asking the illiterate farm laborer living on a dollar a day if broadband is her priority for public expenditure.

Certainly, evidence from across the OECD suggests that the market can take care of business broadband needs without support. Member countries saw the coverage of greater bandwidth digital subscriber lines increase from 33 to 76 percent of lines from 2000 to 2003. From 1999 to 2003, the number of subscribers to broadband increased from 3 million to 82 million. That is phenomenally rapid—it took mobile services twice as long to achieve the same result. Given the speed of rollout, a Working Party of the OECD set up to study this issue concludes that the "main message for OECD policy makers is to give the market time to develop broadband access." The most effective policy tool, they argue, is competition, which "is likely to be far more conducive to the rollout of broadband availability than funding in the form of subsidies."[20]

At the same time, municipal government attempts to set up their own broadband networks have floundered, even in the United States, where 50 percent of initiatives are unlikely ever to break even.[21] At the US state level, universal service policies and tax breaks aimed at areas "underserved" by broadband have no correlation with broadband penetration rates.[22] That suggests that the new technology provides little excuse for governments to reprise their role as telecommunications service providers.

In short, to a large extent, the corporate digital divide in terms of advanced use is likely to shrink without direct intervention, if governments follow policies that support an improvement in the broader business climate. The most constructive way to use the Internet for development may well be to push development to allow for more widespread use of the Internet.

Regarding tax breaks and tariff reductions for software and hardware imports, the negative impact of baroque networks on productivity might suggest a good reason to tax software at very high rates, especially for early adopters. Even if the decision makes financial sense for these early adopters, there is a negative network externality attached to such adoption because it forces other companies to buy new software and hardware they do not need from the perspective of their own business processes, merely so that they can keep in contact with early adopters. The early adopters should be forced to take account of this negative externality when they buy networked software.

Three "no regrets" policies will help to improve Internet access and use, among both individuals and enterprises in developing countries. First, reform of the basic telecommunications sector not only improves access to telephony but also increases Internet use.[23] Carefully designed access programs may be worth supporting to improve access to telephones. Second, some legislation is also necessary to create the enabling environment for e-commerce—and removing other legislation that inhibits Internet use can also provide "quick wins." Third, the Internet also has a number of uses in improving government operations, especially in the area of tax and budget management. Governments that effectively use the Internet to improve the efficiency of internal functioning can provide a range of Internet applications to encourage more widespread Internet use in the business community.

One study found that between 50 and 70 percent of the costs of Internet access in Africa—including the cost of the computer—were accounted for by information infrastructure charges, primarily charges related to the telephone company. The basic infrastructure of wires, waves, and fibers is also clearly responsible for the quality and quantity of Internet access a consumer can expect.[24] Given that, anything that might increase the reach and quality of information infrastructure while reducing the cost is clearly going to be valuable in spreading Internet use. Indeed, it appears that the level of Internet use in a country is even more closely associated with the spread of the basic telecommunications network than it is with income levels.[25]

One can argue about how much a process of reform involving the creation of strongly regulated private competition will improve delivery of telecommunications services. The great majority—up to 95 percent—of the variation across countries over time can be explained by just the two factors of technological advance and country income. Nonetheless, it appears that telecommunications sector reform to introduce private competition does work. Across countries, reform is associated with a 21 percent higher level of labor productivity and an 8 percent higher level of mainline penetration.[26] It is also associated with considerably higher mobile penetration.[27] Reform involves not just the first step of introducing private competition, but also the second step of designing (the least possible level of) regulatory interventions to ensure there is no abuse of market power, especially in areas such as interconnection.[28] Allowing competition, even in an unregulated regime, can provide a better alternative than unnecessary intervention through ownership or limits on competition, as the case of Somalia suggests. In a country where the central government is based in a neighboring state, there is no regulator, and teledensity reached zero in 1991, the sector has seen the emergence of nine private operators providing services to every province, city, and major town, fixed teledensity that is three times neighboring Ethiopia's, and international call costs that at sixty cents a minute are some of the lowest in Africa.[29]

One particular area where deregulation reform can provide a short-term boost to public Internet provision is through the deregulation of Voice-over Internet Protocol (VoIP). If people are able to use public Internet terminals to make voice calls to friends and relatives overseas, bypassing traditional international operators, they can significantly reduce the cost of their call while providing an important revenue stream to telecenter operators, especially where there are still limits on international voice competition.

Beyond regulated competition, in some countries, the government may support minimal rollout of the telecommunications network to areas that remain unserved after pro-competitive reform—primarily to provide voice services that the poor actually widely use, but with the added benefit of extending potential Internet access. The Chilean reverse subsidy auction scheme provides a mechanism to subsidize the private provision of access beyond the market in a manner that keeps the cost of that provision to a minimum. Chile has achieved near-universal telephone access by first opening up the sector to private competition and then auctioning subsidies to the lowest bidder to provide public telephones in unserved areas of the country, at a cost of a little under $10 per newly served citizen.[30]

A recent estimate suggests that replicating the Chilean model worldwide, using the technology and rollout levels of the 1990s, would cost in the region of $5.7 billion. Because of advances in technology and the massive rollout of unsubsidized services, that estimate is considerably higher than the true cost of global universal access in 2005. Indeed, even in large parts of Africa, mobile operators claim it is only a matter of time before they reach near 100 percent footprint coverage based on commercial logic alone. Nonetheless, subsidy costs in some parts of the world might still be significant—Sudan, Democratic Republic of Congo, and Ethiopia alone may require over half a billion dollars in subsidy.[31]

As noted above, extending similar subsidies to telecenters in rural areas is far more expensive than extending telephony and is likely to carry significantly lower cost-benefit ratios. In some cases, there may be a role for government support of backbone and Internet point of presence capacity using the same general approach, however.

Beyond infrastructure rollout, e-commerce requires a supportive legal framework in the banking and industrial sectors, as well as legal and juridical changes in response to challenges that have emerged in tandem with the new technologies. These include standards and protection of digital signatures, the liability of value-added networks, regulation of certification authority, protection of intellectual property, and computer crime and data protection.

To take the example of taxation, the EU has decided to tax goods bought over the Internet—including digital goods such as software downloaded over the web. (That has international implications. To comply, non-EU businesses must either set up headquarters in an EU state and pay that country's value-added tax [VAT] or pay VAT at the rate current in the consumer's home country.) Conversely, in the United States, the Internet Tax Freedom Act and follow-on legislation, combined with US Supreme Court decisions that deny the responsibility of out-of-state suppliers to collect state sales tax, has made e-commerce in the United States effectively tax-free. Estimates suggest that if online transactions carried the same taxes as physical purchases, US e-commerce revenues would be 24 percent lower.[32]

From a theoretical perspective, it is difficult to see a justification for exempting online sales from taxes. Indeed, especially in developing countries where the Internet (and credit cards and logistics services) are largely the domain of the wealthy, tax-exemption would be highly regressive. At the same time, it should be noted that enforcing tax payments across borders for digitized goods in particular may be complex (the EU has apparently persuaded major e-commerce sites like eBay and Amazon to comply with its rules, but the same might not apply for a small African state). However, that is currently a small problem—trade in digitizable goods accounted for less than 1 percent of total global trade in 2000.[33]

A particularly complex set of issues surround legislation on privacy that will have more immediate cross-border effects. For example, the EU threatens a "data embargo" on the export of personal information about its citizens to countries that do not have legislation guaranteeing privacy protections. At the same time, a number of developing countries' concerns with security have led to laws like those in Singapore and Malaysia that force disclosure of encryption keys.[34] How much such conflicts will frustrate LDC company efforts to export using the Internet is still open to question.

In addition to passing new legislation to cover issues such as balanced taxation of e-commerce, countries would also benefit from removing legislation harmful to sector expansion. Marcus Franda provides a list of sector-specific Internet legislation from countries worldwide that includes censorship, registration for users, limiting encryption, restricting licenses for use, requiring ISPs to route traffic through government intelligence systems, holding ISPs responsible for all Internet content, or in extreme cases enforcing a government monopoly over the entire sector.[35] In addition, a number of nonsector laws can significantly damage sector growth, such as rules on the extent of foreign direct investment (FDI), excessive licensing, or overgenerous employment law. Finally, the sector is frequently treated as a cash cow, with additional, sector-specific taxes. In Kenya and Uganda, the sector faces a 10 percent excise tax on top of normal business income taxes. In Afghanistan, various taxes on telecommunications accounted for 14 percent of domestic revenue in 2005, and in Turkey, taxes account for half of the price of a new mobile phone connection.[36] Although these taxes do not appear to have slowed the sector's phenomenal growth to date, they will undoubtedly have a negative impact on providing services to the marginal customer.

Policies for Government Use in LDCs

Turning to the state's use of networked computers to improve the delivery of its own services, we have seen that some applications might deliver significant returns—budgeting, transparency and e-procurement being three clear cases. The government can act as a significant source of demand for the development of a local ICT industry. There is also a potential role for government as a catalyst of local applications, perhaps through support of Internet applications for government-to-business interactions. But it is doubtful that the Internet will become a powerful tool for reaching the average domestic consumer in the near term. And because of weaknesses in infrastructure and institutions, the Internet has lower utility for LDC governments, just as it does for LDC consumers.

At the same time, the cost and complexity of computerization cries out for cautious implementation. A recent project in just one Indian state (Andhra Pradesh), designed to computerize revenue offices, involved training 5,000 staff members and installing 4,500 computers in 1,124 sites over a quarter of a million square miles to handle a database containing over 80 million records. Multiplying these numbers up to the national level in a country as large as India and across a wider range of functions, the scale of the process becomes clear. All this suggests that, although a range of high-return investments is possibie in low-income countries to improve government performance through ICTs, programs should be incremental, carefully designed, and evaluated between each step.

ICT programs should also ensure integration with initiatives at the ministerial and local level. We looked earlier at the Standish Group's 1994 survey of IT executive managers that found that only 16 percent of IT projects were completed on time, on budget, and with all features and functions as originally specified (and the record is even worse in developing countries). The primary reason for the low success rate was given as low user involvement, with limited management support and incomplete requirements and specifications coming in second and third. It is clear that a top-down policy of revolutionary IT innovation is far more likely to suffer from such failures than an organic approach building on reform efforts within government departments.

That suggests a number of lessons. First, the introduction of ICTs should be part of a broader reform process in which the acceptance of an agenda for reform predates the development of a potential róle for solutions involving advanced information systems.[37] It is also very likely that in environments with limited technical capacity, where training and support costs for computers can alone add up to as much as five or ten times the cost of equipment, revolutionary reforms will fail. Evolutionary processes involving carefully piloted projects and stakeholder involvement have to be preferred.

Given the cost and complexity of introducing ICTs, prioritization is also very important, and it is likely that back-office functions such as the processing of tax and land records will have the greatest impact, rather than direct contact with service consumers. For example, attempts to roll out computer labs across schools in poorer developing countries would swamp available budgets. Using computers as part of an effort to track school budgets, staff, and equipment, however, may foster considerable returns.

The importance of an evolutionary approach based on contemporaneous reform and back-office systems suggests that creating a high-profile "e-czar" to push ICT use across government may be risky. However, the e-czar or e-minister should be involved in coordination and regulation to ensure interoperability, rather than the creation of schemes to place all of government on the Net—decentralization and submission of Internet use to broader sector aims should be the guiding principles.

The Role of Donors

If broad-based development is the goal of donors, they should be supporting interventions that will provide the highest possible returns and have an impact on the greatest number—much like national governments. Such interventions occur when people and corporations acting alone cannot provide a good or service that will have a significant impact on the quality of life—including profitable employment. In other words, interventions should happen in response to "market failure." And, as with national governments, interventions should sustainably increase the incomes (or, more broadly, the quality of life) of everyone but should have an especially large impact on the poor. For the same reasons that there are grounds for caution in government support for ICT ventures (because this support is unlikely to be pro-poor or market-restoring), donors also should be wary of supporting the ICT industry, influencing businesses to move online, or setting up Internet-enabled telekiosks in remote rural areas.

The development community has spent a good deal of time and not inconsiderable resources in "pilot projects" related to the use of ICT in development. In an environment in which technology is changing so fast, new opportunities for exploiting those technologies emerge all the time, and our knowledge about what will actually work in the field is so limited that pilot projects are vital. Having said that, pilots need to be designed with care. In particular, they should truly be innovative, should tackle a real development problem that requires public intervention, should have the clear capacity to be expanded nationwide—or at least far more widely—if they do show potential, and should be monitored and evaluated so that we can learn from them. It is an understatement to say many pilots funded by the development community fail to meet some of these requirements. Many ICT projects never get beyond pilot status. They are not designed with sustainability in mind, their demand for upfront donor or government resources would dwarf potential sources of financing if rolled out countrywide, and they are implemented with little effort to discover if local communities actually want the services provided.

The Sustainable Access in Rural India (SARI) project in the Madurai district of Tamil Nadu is an example of a well-designed pilot exercise rolling out ICT access (including the Internet) to fifty villages. The project is a collaborative venture between the Indian Institute of Technology at Madras, the N-Gyan foundation, and groups at Harvard University and MIT. It uses a wireless local loop technology developed at the Indian Institute of Technology at Madras to provide voice and data connectivity to previously unconnected villages at a cost of about $320 per line. Eighty telekiosks now have the technology to provide basic Internet and voice communications at a break-even cost for equipment and connectivity of about $3 per day per telecenter—suggesting that the model is replicable even in fairly poor communities. By ensuring that local operators can provide useful services (three major uses are e-mail to family members working in Persian Gulf states, finding astrological data, and watching movies), SARI's facilities have been used by 23 percent of the target population. Monitoring and evaluation has included frequent surveys of users and an analysis of replicability.[38]

It should be noted that the SARI project does not demonstrate the utility or feasibility of universal Internet access (and the project designers do not claim that it has). Only a quarter of the telecenters were breaking even in 2003, not allowing for the need for operators to make a living, and according to the project designers, the break-even point would be at least twice as high in less favorable environments—Madurai is flat, population-dense, on the electricity grid, and close to a fiber-optic backbone. Nonetheless, the project does meet the tests for innovation, demand, replicability, and strong monitoring and evaluation.

Certainly, donors can support technical assistance in areas such as regulation and policy development (although perhaps not a second or third generation of "e-readiness" studies). Donors can also support investment in potentially sustainable methods of improving poor people's access to information—radio, the educational use of television, or perhaps basic public access to telephony—through reverse-auction methods. But contrast most telecenter projects with community radio projects. Community radio costs are a fraction of those for a telecenter project, as we have seen. Radio's reach is also greater—the illiterate and those distant from the station can listen. And such community stations are sustainable. Donor-funded telecenters to date have largely failed the usage, impact, and sustainability tests. Only in exceptional circumstances is it likely that donor support for widespread government-owned or even government-subsidized rollout of public access Internet services is the most efficient use of resources—worth diverting funds from basic health programs, for example.

The largest impact of developed countries on the use and utility of the Internet in the developing world will not come through aid, however. Much as OECD trade barriers cost developing countries more than total aid flows return, aid directed at ICT use is dwarfed by disincentives and costs that rich country policies have placed on the successful use of ICTs by the developing world.

One major issue that divides wealthy and poor countries is the method of apportioning the costs and revenues of the international flow of data and voice over the information infrastructure. Because of the somewhat archaic rules of international telecommunications agreements, voice traffic is still treated somewhat differently than data traffic in this regard, but in both cases, the developing world claims that it has been unfairly treated, perhaps with some justification.

The "traditional" revenue and cost-sharing arrangements covering voice circuits are governed by the international accounting rates regime, under the auspices of the International Telecommunications Union (ITU). In this system, two countries agree on the rate that they will charge each other to complete a call from one country to the other. (For example, a minute-long call from the United States to Chad will be completed by Chad at a charge to the United States of $2, and vice versa.)

The system is struggling under the influence of international competition and alternate methods of calling (over the Internet, for example)—especially because accounting rates have traditionally been set far above the cost of completing a call (creating strong incentives for bypass). Some countries have agreed on a bilateral level to abandon the accounting rates regime altogether and just allow their private competitive phone companies to set their own termination agreements. But what has upset some developing countries in recent years was a unilateral decision in 1998 by the US Federal Communications Commission (FCC) to set upper limits on accounting rates at levels that it estimated reflected the approximate cost of terminating a call—rates that at below twenty cents a minute reflected average rate drops of 60 percent. The FCC acted in part because the United States was paying out $4.5 billion a year in accounting rates settlements because of an imbalance in traffic (more out than in).[39]

Developing countries have argued that they used the revenue from this US "trade deficit" in telecommunications to fund rollout of networks. In fact, the evidence shows that countries that received the highest payments did not see more rapid growth of their networks, probably because the money was siphoned off by national treasuries.[40] Furthermore, the accounting rates system was not designed to act as an international cross-subsidizing mechanism for telephone users in rich countries to support rollout of services in poor countries, and it would have worked poorly if this had been the intent. For example, only 2.5 percent of the US outpayments went to sub-Saharan Africa—the region with perhaps the greatest need for such support.[41]

But even if the accounting rates debate boils down to a good deal of fuss over nothing much, there is still some justification for the case made by LDCs, at least regarding access. US consumers do benefit from new connections being made in LDCs because of the famed "network externality" nature of telephony, and so some sort of subsidy for providing such access might be justified. The scale of such a subsidy and how it would be fairly apportioned is open to question, however, and it may be beyond the ability of the international system to design and enforce one. Certainly, the decision of the recent World Summit on the Information Society to "study" rather than set up an international fund to support information infrastructure rollout suggests the inherent political minefields.

Perhaps a more valid concern regards global Internet traffic and standards. At the moment, LDCs pay for the entire cost of the data connection to the Internet in a developed country (usually the United States or the UK). So, if a man sitting in Nairobi sends an e-mail to a woman in Washington, D.C., the e-mail travels along bandwidth paid for by his African ISP until it reaches the United States—all the way over the international satellite or cable connection. When the woman replies, once the e-mail reaches the international connection on the coast of the United States, the African ISP pays for its transit again. The nature of the Internet makes it difficult to calculate what a "fair" apportionment of costs would be. Who benefits from the African downloading material? Is it spam or requested content originating from the United States? It is difficult to know who should pay. According to one Kenyan ISP, the "Africa pays all" system costs African Internet users $500 million a year.[42] That claim may be exaggerated, and in common with the accounting rate regime, there may not be an implementable regime that would be more just. Nonetheless, developing countries may have a case in suggesting that the current regime is skewed against them.

There are also a number of international governance issues that matter regarding the "soft" side of the Internet (as opposed to the infrastructural issues discussed above). In particular, there are issues surrounding patents and copyrights that allow monopoly profits to companies such as Microsoft while costing developing countries in particular considerable resources in terms of hardware obsolescence. The trend toward stronger enforcement of patents over a wider range of "products" (such as one-click shopping) only exacerbates such issues.

The problems caused by current copyright and patent law are perhaps best illustrated by the fact that Mickey Mouse is still under copyright protection, thanks to a recently granted extension of copyright. There is little economic justification for such extensions (surely Mr. Disney's incentive to design Mickey Mouse was little influenced by the fact that his copyright would be extended forty years after he died). But there are good reasons to think such copyright extensions limit spillover benefits of innovation to the broader economy, in particular the developing world. Yet such protections are being strengthened at the international level. For example, the agreement on Trade-Related Aspects of Intellectual Property Rights, part of the last WTO trade round, required all member countries to protect copyrights, patents, and trademarks for at least twenty years. Prior to TRIPS, the average length in developing countries was four to seven years. Because LDCs are net importers of R&D-intensive products, the expansion of such rights, at a point in their development far earlier than similar rights were adopted in the West (Spain only met the standard in 1992, for example) has significant short-term costs.[43]

As UNCTAD noted in its recent policy discussion paper on intellectual property rights, some level of such rights is necessary both to attract creative and technology-intensive industries and to ease the willingness of such industries based overseas to provide products. Nonetheless, recent research provides little evidence of a downside to limited IPR regimes in the form of lower high-technology imports or flows of FDI.[44] IPRs do carry obvious costs, however, in terms of higher prices for goods such as computer programs currently frequently copied or produced without license.[45] Implementing TRIPS would carry immediate payments from patents and copyrights of $20 billion from developing to developed countries each year.[46]

This issue extends far beyond the Internet and ICTs, obviously. But because of the negative externalities attached to the development of baroque software, it has particularly severe implications in the ICT sector. Perhaps the most important thing that donor countries could do to overcome the digital divide would be to reconsider pressures on developing countries to conform to an intellectual property system that may be of little overall economic benefit at home or, especially, abroad.

To limit the damage that OECD patent and copyright holders in the software arena may be doing to both the late-adopter companies in the United States and LDCs, perhaps operating system monopolies should be regulated: fair licensing enforced, predatory pricing controlled, and open source software encouraged. Going further, new versions of old software have a negative externality attached—perhaps they should be significantly taxed, or at the very least should not attract subsidy and tax breaks.

To conclude, for developing country governments, the overarching policy lesson is "proceed with caution." There are opportunities to be grasped in both the production and use of Internet applications and infrastructure. The theory and empirics of the role for government support are murky at best, however. Widespread subsidy of either production or use does not appear to be justified by the evidence. Government use of the Internet to improve its own operations has potential but also carries great risks. An evolutionary approach beginning with back-office functions and tied closely to large reform efforts is perhaps most appropriate. For donor countries, revisiting international governance of infrastructure and intellectual property rights is likely to have a far larger impact on the sustainable use of the Internet in developing countries than are programs to roll out rural telecenters.

Notes

1. The prime example is perhaps the $4 billion Ajaokuta Steel plant in Nigeria, which had not poured metal even twelve years after construction (see Pritchett 2002).
2. The data in this paragraph are culled from a number of sources: US air conditioner ownership from Myers 2002; Indian air conditioner ownership from Gallup Organization 1996; and Internet and income statistics from World Bank 2001a. The link between air conditioning and productivity has been convincingly made by Lee Kuan Yew with regard to Singapore (see http://www.neec.gov.sg/aboutus/sub_speech06.shtm) and by Raymond Aresnault with regard to the southern United States (see Aresnault 1984).
3. Dedrick and Kraemer 1998; Matthews and Cho 2000.
4. Kumar and Joseph 2005.
5. Kolko 2002.
6. Hobday 2001.
7. Ostrov 2002.
8. Dedrick and Kraemer 2000; World Bank 2004.
9. See Madani 2000. The worrying news is that Rwanda is already following such a strategy, with an ICT commission headed by its president and a national ICT policy for the country that sets up a top-level national IT agency to oversee a 400-page, five-year, US$500 million plan and strategy for ICT. The country is far from alone—even Iran's latest five-year plan emphasizes ICT.
10. World Bank 2001d.
11. Richardson and Gillespie 2003.
12. UNCTAD 2003a.
13. Srinivasan 2005.

14. Oxfam 2000.
15. Pyramid Research 2001.
16. Eltzroth and Kenny 2003.
17. Eltzroth and Kenny 2003.
18. World Bank 2004.
19. World Bank Regional Program on Enterprise Development at rru.worldbank.org/EnterpriseSurveys.
20. OECD Working Party on Telecommunications and Information Services Policies 2004.
21. Jupiter Research 2005.
22. Wallsten 2005.
23. Wheeler et al. 2001; Clarke 2001; Kenny et al. 2005.
24. Africa Internet Forum 1999.
25. Wheeler et al. 2001.
26. Fink; Mattoo, and Rathindran 2002.
27. Keremane and Kenny 2005.
28. Wellenius 2005.
29. World Bank 2005.
30. Kenny 2002a.
31. Keremane and Kenny 2005.
32. Goolsbee 2000.
33. Mattoo and Schuknecht 2000.
34. Mann, Eckert, and Knight 2000.
35. Franda 2002.
36. *Economist,* July 8, 2005.
37. Heeks 1998.
38. Best 2002.
39. Braga, Forestier, and Stern 1999.
40. Wallsten 1999.
41. Braga, Forestier, and Stern 1999.
42. "The Great African Internet Robbery," bbc.co.uk, April 15, 2002.
43. Subramanian 2004.
44. Fink and Maskus 2005.
45. UNCTAD 2001.
46. World Bank 2002a.

7 —

Confronting the Costs

DAVID NOBLE, A PROFESSOR of history at York University in Toronto, complains that, over the past few hundred years, we have made technology our new belief system. Because of the "religion of technology . . . we routinely expect far more from our artificial contrivances than mere convenience, comfort, or even survival. We demand deliverance."

Communications technologies, perhaps unsurprisingly, have played a particularly prominent role in technological utopianism. The *Communist Manifesto* itself proclaimed that the railway would engender revolution in a matter of years. The "union of the workers . . . is helped on by the improved means of communication that are created by Modern Industry, and that place the workers of different localities in contact with one another . . . to attain which the burghers of the Middle Ages, with their miserable highways, required centuries, the modern proletarian, thanks to railways, achieve in a few years."

Noble sees the networked computing industry as particularly rife with techno-utopian visionaries who predict the obsolescence of the physical body and immortality through existence in cyberspace (one wonders who will be left around to reboot the network).[1] Books such as Michael Dertouzos's *What Will Be* or Esther Dyson's *Release 2.0* may use the relatively new technology of the networked computer as a basis for their visions of a better tomorrow, but the formula is clearly an old one.

At the same time, Danny Quah, a professor at the London School of Economics, takes a well-aimed swipe at the Internet skeptics in the economics profession, based in part on the compelling evidence that there *has* been a significant change in work and play habits as a result of the Internet:

> Pick up a newspaper today, and you have to realize how words and concepts that didn't even exist a decade ago—Internet browsers, desk-top operating

> systems, Open Source Software . . . now appear regularly in front page headlines. These headlines describe *news* items—not science fiction trends. . . . When you live in that world, it is puzzling when you meet people intent on proving to you that none of those things you think you see and experience is real. . . . These skeptics show you charts and figures, bristling with numerical calculations, arguing that the changes you figured to be deep and fundamental apply, in reality, only to the miniscule group of people working in companies that manufacture computers.[2]

Quah argues that we are looking for the impact of the Internet in the wrong place when we study productivity statistics and other supply-side variables. We should be looking at how the Internet enables consumers to consume an ever-greater quantity of "knowledge goods"—writing, music, films, and websites with pictures of dogs' noses.[3] That is where the impact is growing and is already large.

Although techno-utopianism is overblown, Quah is surely right that in OECD countries, the big impact of the Internet is clearly on consumers, in ways that are poorly reflected in the usual raft of economic statistics. But even here there are reasons to think that not all the impacts have been positive, and the changes (again) might not be as large as they first appear.[4]

Take the sociological aspects of work. The new technology has allowed new working relationships to blossom, and it has allowed (some) people to work, and enjoy working, from home. At the same time, the great majority of the less than 6 percent who "work from home" remain in the sadly preindustrial sectors of child care and farming. Even among those who try to work from the telecottage, 50 percent give up within the year and go back to the office. And anybody who believes that the Internet will free the world of tedious, soul-destroying jobs should read Mike Daisey's description of working at Amazon.com's customer service division, *21 Dog Years:*

> On weekends I loved Amazon and I would speak at great and windy length . . . "God, I love my company, I love working, it's so great, we're making history." . . . But when I came into work I flipped—while there I hated the place, hated the phone and the email and the endless tracking. Most of all I hated meeting with my supervisor, because at Amazon there was never a positive reason to have that meeting—it is always punitive when metrics exist to show where to bring the hammer down.[5]

Apparently, loathed bosses and lifeless jobs did not go out with the end of the industrial age.

Or take the social revolution suggested by Dertouzos:

> As she comes close, her bodynet handshakes with yours, and you are embarrassed to see on your glasses that you are the one with the inferior model. . . . As you ponder the need to upgrade your system, the two bodynets exchange

> your personal auto-profiles. . . . You now see in your glasses that . . . you share a common interest in "early music." Seeing a similar message, the woman smiles appreciatively and lets her bodynet burst to yours her favorite Italian madrigal.[6]

This is just before she walks into a lamppost, addled by "handshaking," glass model superiority, reading those glasses, and madrigal file selection all at the same time. The point here is not that such equipment can't be invented. Indeed, Dertouzos pointed out that most of the technology already exists. It is that much of the equipment suggested won't be invented because people are so ill-designed to use it. People, culture, institutions—not invented technologies in a vacuum—play the greatest role in setting the rate of "social progress," just as much as they set the pace of economic progress.

A number of political scientists and popularizers early declared that the Internet would spread democracy around the world, including Joseph Nye of Harvard and Thomas Friedman of the *New York Times*. These ideas were taken up by aid groups such as the National Democratic Institute, which started putting the Internet in libraries in developing countries, for example. And a number of governance initiatives have been supported by the Internet. In South Africa, the government used the Internet to encourage citizen input in drafting the postapartheid constitution. The Constitutional Assembly created a website that made available draft texts, political party positions, and committee reports and recommendations. In addition, citizen feedback via e-mail was encouraged, and a virtual community of individuals and groups was able to share ideas on proposed drafts of the constitution.[7]

But as Joshua Kurlantzick of the *New Republic* points out, the overall impact of the Internet has been less clear-cut—Singapore, always high on "e-leadership" rankings, saw the ruling party increase its share of the vote to over 95 percent at the last election. Malaysia, another e-leader, has also seen the entrenched party strengthen its grip. As John Daly has pointed out, the Internet might lead to more democracy, but it might also improve the environment for kleptocrats, plutocrats, bureaucrats, theocrats, or aristocrats.[8] As with economic performance, estimates of the Internet's impact on political realities often forget the fact that ICTs are a tool, their impact dependent on a range of environmental factors. For example, does the government monitor, screen, and/or block content? An added factor is that the Internet, unlike broadcast technology, atomizes individuals rather than bringing them together. Web surfing is a lone activity that involves interactions with unknown people who can pretend to be what they are not, as children are constantly warned. That is not exactly the safest environment in which to collect together people who share discontent with an oppressive regime, as Winston Smith discovered to his cost in *1984*.

Just as utopian visions have been based on computers and the Internet, so have dystopias. Ben Seligman, as early as 1966, argued in *Most Notorious*

Victory that computer-driven automation would throw millions out of work and "threatens to destroy the essential human qualities that have been thus far characteristic of man."[9] Orwellians can comfort themselves with the idea of permanent government vigilance through the Net. In a decidedly downside article for *Upside* magazine entitled "Forget Digital Utopia. . . . We Could Be Headed for Technofascism," columnist Michael Malone notes that libertarians of the right, including George Gilder and Steve Forbes, are uniting in a strange alliance with techno-utopians of the left such as Al Gore and the writers of *Slate*, to suggest that all is for the best in the best of all possible networks. Yet he argues: "Does any reader of this article intuitively sense that everything is getting better? . . . Thanks for the fetal monitors and anti-lock brakes, but can you please take back the alienation and the porn spam-mail. . . ."[10] Karl Marx (yet again) could have been presenting this view of the Internet (rather than making a more general point about technology) when he wrote: "The new-fangled sources of wealth, by some weird spell, are turned into sources of want. . . . At the same pace that mankind masters nature, man seems to become enslaved to other men or to his own infamy. . . . All our invention and progress seem to result in endowing material forces with intellectual life, and stultifying human life into material force."[11] It seems that the Internet is truly everything to everybody. Whatever our previous hunches about the way the world is going, the Internet allows us to think it will get there faster.

As a group of general purpose technologies, ICTs will also have a widespread impact across sectors connected with environmental change. The "weightless economy" is clearly less polluting than the weighty one. If the Internet leads to the substitution of web browsing for driving the SUV to the demolition derby, it will reduce the level of pollution per dollar of GDP. Furthermore, the Internet can be used to improve the delivery of environmental services, such as tracking migratory animals. But precisely because ICTs are general purpose technologies, they can also be used for the greater production of environmental harm, such as getting dirty coal to the power plant more rapidly, advertising rhino horn aphrodisiacs, or using global positioning systems and satellite data to increase unsustainable fish catch. The list goes on. And add to that all the chemical waste produced in manufacturing computers.[12] Absent intervention, there's no particular reason to think that investments in ICT will be a net positive for the environment, even while average investments in ICTs may be less (directly) environmentally harmful than investments in off-road vehicles or biological weapons factories.

Danny Quah has a point, then—the short-term social impact and, potentially, the long-term economic impact of the Internet is missed by a focus on productivity statistics. But it is likely that *both* the positive *and* negative social and environmental consequences of the technology are missed by such an approach.

From the standpoint of this book, the more important thing to note about Quah's argument regarding the significant impact of the Internet is that it is a rich world argument. There are more Internet users in developing than developed countries, but nonetheless the great majority of people are not Internet users, and that state of affairs is unlikely to change any time soon. Thus the primary impact of the Internet (if there is to be one) will be on the supply side. And in the developing world, where there are not enough old-fashioned solid goods, where labor productivity really is too low to guarantee a minimum standard quality of life, those supply-side statistics matter.

Reviewing the numerous case studies from India reported in earlier chapters might help summarize the case for caution regarding the Internet as a force for development. Information and communications technologies, including the Internet, have had a significant impact. Software and outsourcing industries in India employ 800,000 people. Numerous public and private networked computing applications are improving service delivery—Bhoomi in Karnataka has made land titling significantly more straightforward, and SARI is providing telecommunications and Internet services in rural areas.

At the same time, the total factor productivity gains attributable to ICT production in India from 1995 to 1999 add up to 0.05 percent of GDP—and much of the benefit from these gains is likely to go to consumers in other countries, rather than to Indian companies and employees. On the usage side, microeconomic studies find little evidence that ICT use increases company productivity in the country. There are a number of examples of failed Internet applications in the public sector that suggest the complexity and risk of moving online. This risk applies especially to attempts to use the Internet to directly connect government with citizens, such as Gyandoot with its handful of users. Even where the Internet is available to poor people, few use it—perhaps 2 percent of those surveyed in villages in Gujarat near public terminals use the facilities, and only a little more than 2 percent of Internet users across India experiment with e-commerce.

These statistics point to a broader truth regarding the use of the Internet—that access is only one part of the equation. Without an enabling environment (reformed institutions of government and business and access to skills, finance, transportation, and so on)—the Internet changes little. Where that enabling environment exists, the Internet can be a useful tool—as we saw with the case of the women's rights group Sakshi, which used the Internet as part of an effort to pursue sexual harassment legislation in India. But most poor people do not live in the enabling environment that makes Internet use attractive, and so, even with access, they do not use the Internet. In part because of these factors, the benefits of widespread Internet access will frequently not cover its costs. To take the case of education, randomized trials in India suggest that benefit-cost ratios for computers in remedial education are one-fifth of those for employing more teachers. Overall, the Internet and IT more generally has had an impact in India, and that impact will grow—but there is little evidence that the Internet is a leapfrogging technology that will allow India to rapidly and sustainably improve overall development outcomes at the macroeconomic level.

The Internet has a role to play in developing countries. If economic growth is about technology, and technology is about applying knowledge, a tool of knowledge transfer must have a potentially significant impact. Each year, the world produces an estimated 2 exabytes of optical, paper, film, and magnetic storable information (or, as the *Economist* noted, the equivalent of 20 billion copies of the *Economist*—which works out to 634 unique copies of the magazine a second). The Internet is making considerably more of this information available to more people worldwide.

The problem, however, is not and never has been lack of information output. It is getting access to the right information input.[13] Rather than a tool that gives us untold information at the click of a button, what we most need is a way to filter information.[14] Take research on the Internet as a force for economic change as an example. We've already seen that the problem with this research, especially in developing countries, is that it is largely made up of unsubstantiated anecdote. What the Internet allows is for those anecdotes to spread far and fast, to be reposted and rewritten hundreds of times. When I typed the words "Internet," "education," and "India" and the phrase "hole in the wall" into Google on February 9, 2006, I got 46,200 results. Over 40,000 different ways to access the same story—a story that (however innocently) probably misleads as much as it informs when it comes to the potential impact of the Internet in LDCs. We're still missing the hard evidence on this topic, Internet or no.

More broadly, a technology that—by providing information unrelated to work facilitates perhaps 172 million hours a week of time wasting in US corporations (as we saw in Chapter 3); a technology that costs ten times in technical support than it costs in hardware; a technology that isn't used at all by a huge majority of the poorest people in the world, people whose total yearly income wouldn't pay for the costs of access and who lack the infrastructure and financial institutions to fully exploit it—may be a strange technology to push as a driver of economic revolution in the West, let alone in the Third World.

To repeat, ICTs have already proven themselves of great utility in developing countries, as demonstrated by their incredibly rapid adoption. In particular, the spread of mobile telephony surely does constitute a revolution, with more than half of the rural population of the developing world now able to talk over distances to friends, family, business partners, and government representatives. And technology has already blurred the distinction between telephony and the Internet, with impacts on usage patterns in the developing world. In Senegal, for example, fishermen are using text-messaging services to alert buyers of landing times and tracking prices. Sustainable Internet use itself will also spread ever further as technologies make the costs of access fall further. Technology has a huge role in a number of government functions—some developing countries suggest that e-procurement systems are saving them 20 to 30 percent on purchases, for example.

However, such uses are frequently marginal improvements over second-best alternatives: Text messages from fishermen are only a marginal improvement over using the phone to call in landing times. For the Internet to be a tool of economic revolution, it needs to be used to accomplish dramatic new things—the eBay model, as it were. And it is precisely these more revolutionary tools that low-income populations tend to be least equipped to use.

Again, this is not to suggest that some companies in the developing world find little use for the Internet as a marketing tool because their customers don't buy much online. But that is a rather insignificant manifestation of a much larger problem. There are many reasons beyond few companies selling them goods that stop consumers using the Internet to buy things—income, literacy, lack of credit systems, and no way to receive the goods might be four obvious ones. As we've seen, there's strong evidence to think that it's not the access problem that's the big one for advanced use of the Internet in developing countries.

In the developed world, the Internet has played an important role in keeping returns on investment up. If, since the agricultural revolution, there had been no new inventions that allowed new investment opportunities, we'd be 10 feet deep in ploughshares—and much poorer. But in the developing world, a lot of older inventions still have the capacity to generate significant investment returns—investments such as turbines, tarmac, and books. Indeed, turbines, tarmac, and books are a prerequisite for successful investment in many uses of the Internet, so it seems probable that more investment should flow toward them.

For policymakers, there are opportunities from the technology of the Internet, there is need for public support of investment in areas that will catalyze Internet use, and there is also need for policy and regulatory reform to ease the deployment of applications. In particular, there is a need for investment in suitable government applications.

At the same time, on the production side, we have seen that the extraordinary benefits accruing to US producers may be difficult to export, and even though the industry creates jobs and financial returns, there is no strong reason to treat it differently from other industries that also create jobs and investment returns. On the access side, what might be more important than universal Internet access? What about universal access to sufficient nutrition? Or universal access to primary education? Or access to sanitation? The list goes on. Proponents of universal Internet access will say that these are false choices, that the Internet is a tool to get education, nutrition, and sanitation. Perhaps, but is it as effective a tool as a teacher, or food, or pipes?

In some cases, the most cost-effective way to improve education prospects in a country may be to stick networked computers in the Ministry of Education, for example. But those cases will emerge from a careful analysis of what is blocking the expansion of educational opportunities in a particular country. Unevaluated assumptions that the Internet will always be the cost-effective way to improve a development outcome will most likely be wrong. There are real trade-offs and a real choice to be made. And given that, we should focus on universal nutrition, education, and sanitation as goals first and universal access to the Internet second (or third).

Usually, when it comes to IT and development, Microsoft is more part of the problem than part of the solution. Nonetheless, chairman Bill Gates, when wearing his philanthropy hat, has a far more positive role to play and a realistic idea of how to play it. His argument about the place of *direct* support to the Internet in development priorities (including those of his foundation) is perhaps a good way to conclude:

> Let's be serious, do people have a clear view of what it means to live on $1 a day? . . . There are things those people need at that level other than technology. . . . About 99 percent of the benefits of having [a PC] come when you've provided reasonable health and literacy to the person who's going to sit down and use it. . . . I am suggesting that if somebody is interested in equity that you wouldn't spend more than 20 percent of your time talking about access to computers, that you'd get back to literacy and health and things like that.[15]

A focus on the basic needs of people in developing countries will unearth a significant role for information and communications—and even some role for advanced ICTs such as the Internet. But to suggest that the Internet is a leapfrog innovation with the capacity to radically alter global economic inequality far overstates its role. In short, the Internet, and particularly the World Wide Web, remain dubious agents of development.

Notes

1. Noble 1999. He cites books such as *The Age of Mind: Transcending the Human Condition Through Robots* or *Beyond Humanity: CyberRevolution and Future Mind.*
2. Quah 2002.
3. www.dognoses.com.
4. Despite the fact that it is not the focus of this book, I would argue briefly that, as with the economic impact of the Internet, evidence is mixed. People are using the Internet a lot. And they are using it to the exclusion of other forms of infotainment—such as books and television. Clearly, the average Internet user thinks that the technology is pretty exciting—and that his or her world is better for its invention. But as with all information and communications technologies, there are better and worse uses for it. For every person who listens to the National Public Radio story on the endangered hairless wombats of Tasmania (to the obvious benefit of mammalkind), 100 are listening to Howard Stern, New York shock jock (to the obvious benefit of the longevity of jokes regarding genitalia). For every Public Broadcasting Service viewer learning about gender bias in the twelfth-century poetry of Languedoc, 100 are watching Home

Improvement reruns. The costs and benefits of past communications revolutions have long been a subject of heated debate, pitching the likes of Neil Postman (author of *Amusing Ourselves to Death*) against the Cato Institute, with William Bennett somewhere in a third corner, and the Internet only joins the list of debating topics. Indeed, one wonders why we would expect the debate over the impact of the Internet to be any different—although supporters do point out that the Internet is different and better than radio and TV because it is interactive and less amenable to content control—while critics point out that it is different and worse because it is interactive and less amenable to content control.

5. Daisey 2002.

6. From the *San Francisco Chronicle*, "Electronic Noses: Smell the Future," review of M. Dertouzos *What Will Be,* by Laura Evenson, Sunday, April 6, 1997.

7. Lal 1999.

8. Daly 2000.

9. Seligman 1966.

10. Malone1998.

11. Marx's speech given on the anniversary of the *Peoples' Paper*, quoted in Berman, Bound, and Machin 1988, 20.

12. Here is a list of solvents and cleaning agents used in just one part—photolithography—of the semiconductor manufacturing process: detergent, isopropyl alcohol, acetone, ethanol, hydrofluoric acid, sulfuric acid, hydrogen peroxide, hydrochloric acid, nitric acid, chromic acid, ammonium hydroxide, hexamethyldisilazane, xylene, celloslove acetate, n-Butyl acetate, ethylbenzene, chloroflourocarbons, cholorotoluene, glycol ethers, and (less threatening) deionized water.

13. Much of that information is visual. For example, the "information" contained in the average American's consumption of television each year alone is about 3,100 gigabytes. If that information had all been plain text, it would be the equivalent of about 7.9 million 200-page books. (Apparently, technology inflation has left the single picture worth 20,000 bytes—perhaps 4,000 words, rather than the traditional 1,000—and we only get to see that picture for one-thirtieth of a second.) Of course, one would have to have a very high opinion of the informational value contained in a single episode of *Survivor* to think it worth approximately 5,000 books. Nonetheless, even if we limit ourselves to just books, global output equals 987,000 books per year, or one every thirty seconds. "Byte Counters: Quantifying Information," *Economist*, October 21, 2000; Berkeley School for Information Systems and Management 2000.

14. We also need a way to increase the cost of disseminating worthless stuff.

15. Gates 2000.

Acronyms

AOL	America Online
APEC	Asia-Pacific Economic Cooperation
B2B	business-to-business
B2C	business-to-consumer
CABECA	capacity building for electronic communication in Africa
CEO	chief executive officer
ECLAC	Economic Commission for Latin America and the Caribbean
EDI	electronic data interchange
FAO	Food and Agriculture Organization
FCC	Federal Communications Commission
FDI	foreign direct investment
G7	Group of Seven
G8	Group of Eight
GATT	General Agreement on Tariffs and Trade
GDP	gross domestic product
GNP	gross national product
ICT/ICTs	information and communication technology/technologies
IDPM	Institute for Development Policy and Management
IDS	Institute of Development Studies
ILO	International Labour Organization
IMF	International Monetary Fund
IPO	initial public offering
ISP	Internet service provider
IT	information technology
ITU	International Telecommunications Union
LDCs	lesser developed/least-developed countries
LINCOS	little intelligent communities

MIT	Massachusetts Institute of Technology
MSC	multimedia supercorridor
NBER	National Bureau of Economic Research
NGO	nongovernmental organization
OECD	Organization for Economic Cooperation and Development
PADIS	Pan African Development Information System
PPP	purchasing power parity
R&D	research and development
SARI	Sustainable Access in Rural India
SMEs	small and medium-size enterprises
TFP	total factor productivity
TRIPS	Trade-Related Aspects of Intellectual Property Rights
UNCTAD	United Nations Conference on Trade and Development
UNDP	United Nations Development Programme
UNECA	United Nations Economic Commission for Africa
UNESCAP	United Nations Economic and Social Commission for Asia and the Pacific
USAID	US Agency for International Development
VAT	value-added tax
VOIP	voice over Internet protocol
VSAT	very small aperture terminal
WHO	World Health Organization
WTO	World Trade Organization

Bibliography

Abrahmson, B. 2000. "Internet Globalization Indicators." *Telecommunications Policy* 24: 69–74.

Acemoglu, D., S. Johnson, and J. Robinson. 2001. "The Colonial Origins of Comparative Development: An Empirical Investigation." *American Economic Review* 91, no. 5: 1369–1401.

Adkins, D. 1999. "Cost and Finance," in A. Dock and J. Helwig, eds., *Interactive Radio Instruction: Impact, Sustainability and Future Directions.* Washington, DC: World Bank.

Africa Development Forum. 1999. *Internet Economic Toolkit for African Policymakers.* Washington, DC: World Bank.

Albouy, Yves. 1999. "Institutional Reform," in E. McCarthy and F. Martin, eds., *Energy and Development Report, 1999: Energy After the Financial Crisis.* Washington, DC: World Bank.

Almasy, E., and R. Wise. 2000. "E-venge of the Incumbents: A Hybrid Model for the Internet Economy." *Ivey Business Journal* 64, no. 5: 16–19.

Altig, D., and P. Rupert. 1999. "Growth and the Internet: Surfing to Prosperity?" *Federal Reserve Bank of Cleveland Economic Commentary,* September 1999.

Amiti, M., and S. Wei. 2004. "Fear of Service Outsourcing: Is It Justified?" NBER Working Paper 10808, Cambridge, MA.

Annamalai, K., and S. Rao. 2003. "What Works: ITC's E-Choupal and Profitable Rural Transformation." Mimeo, World Resources Institute, Washington, DC.

APEC (Asia-Pacific Economic Cooperation). 2001. *The New Economy and APEC,* Singapore: APEC.

Aresnault, R. 1984. "The Cooling of the South." *Wilson Quarterly*, Summer.

Association for the Development of Education in Africa. 1999. *Prospective, Stocktaking, Review of Education in Africa: Draft Synthesis Document.* http://www.adeanet.org.

Autor, D., L. Katz, and A. Kreuger. 1998. "Computing Inequality: Have Computers Changed the Labor Market?" *Quarterly Journal of Economics* 113, no. 4: 1169–1214.

Baily, M. 2002. "The New Economy: Post-Mortem or Second Wind?" *Journal of Economic Perspectives* 16, no. 2: 3–22.

Baker, W., E. Liun, M. Marn, and C. Zawada. 2001. "Getting Prices Right on the Web." *McKinsey Quarterly* 2.

Balit, S. 1998. *Listening to Farmers: Communication for Participation and Change in Latin America.* Mimeo, FAO, Rome.

Bampo, Gloria. 2001. "Bringing the Internet to Ghana." Submission to infoDev/IICD Stories Project Website at http://www.iicd.com.

Banerjee, A., S. Cole, E. Duflo, and L. Linden. 2004. "Remedying Education: Evidence from Two Randomized Experiments in India." Mimeo, MIT, Boston.

Bank of England. 1999. *Inflation Report,* Summer. London: Bank of England.

Bedi, A. 1999. "The Role of Information and Communication Technologies in Economic Development: A Partial Survey." ZEF Discussion Papers on Development Policy, No. 7, Bonn.

Bellamy, E. 2000. *Looking Backward.* New York: Signet Classic.

Berman, E., J. Bound, and S. Machin. 1998. "Implications of Skill-Based Technological Change: International Evidence." *Quarterly Journal of Economics* 113, no. 4: 1245–1280.

Berman, M. 1988. *All That is Solid Melts into Air: The Experience of Modernity.* New York: Penguin.

Best, M. 2003. "The Wireless Revolution and Universal Access in ITU," in M. Best, ed., *Trends in Telecommunications Reform* Geneva: ITU.

Bhatnagar, S. 2000. *Information and Communication Technology in Development: Cases from India.* Thousand Oaks California: Sage.

Borghans, L., and B. ter Weel. 2000. "Do We Need Computer Skills to Use a Computer? Evidence from the UK." Mimeo, Maastricht University, Maastricht.

Braga, C., E. Forestier, and P. Stern. 1999. "Developing Countries and Accounting Rates Reform: A Technological and Regulatory El Niño?" *Public Policy Journal,* 173.

Bresnehan, T., E. Brynjolfsson, and L. Hitt. 1998. "Information Technology, Workplace Organization, and the Demand for Skilled Labor: Firm Level Evidence." *Quarterly Journal of Economics,* February, 339–376.

Brown, J. S., and P. Duguid. 2000. *The Social Life of Information.* Boston: Harvard Business School Press.

Brown, K. 2000. "Hedonic Price Indexes and the Distribution of Buyers Across the Product Space: An Application to Mainframe Computers." *Applied Economics* 32: 1801–1808.

Business Software Alliance. 2000. *Global Software Piracy Study.* Washington, DC: Business Software Alliance.

CABECA. 1998. "Connectivity in Africa: Use, Benefits and Constraints of Electronic Communication: Synthesis Report Part 2: Overview of the Findings of the Project." Mimeo, Addis Ababa: UNECA/PADIS.

Cassen, R., et al. 1994. *Does Aid Work?* Oxford: Oxford University Press.

Cassidy, J. 2002. *Dot.Con: The Greatest Story Ever Sold.* New York: HarperCollins.

———. 2000. "Do Computers Make You More Productive?" *New Yorker,* November 27, 2000.

Castles, S. 2000. "The Impacts of Emigration on Countries of Origin," in S. Yusuf, S. Evenett, and W. Wu, eds., *Local Dynamics in an Era of Globalization.* New York: Oxford University Press.

Cawthera, A. 2001. "Computers in Secondary Schools in Developing Countries: Costs and Other Issues." Mimeo, WorldLinks, Washington, DC.

Cecchini, S. 2002. "Evaluation of Gyandoot and Bhoomi." Mimeo, World Bank, Washington, DC.

Center for Electronic Governance, Indian Institute of Management. 2002. *Gyandoot: Rural Cybercafes on Intranet: Dahr, Madhya Pradesh, India*. Ahmedabad.

Choudhury, S., and S. Wolf. 2002. "Use of ICTs and Economic Performance of Small- and Medium-Scale Enterprises in East Africa." Paper presented at the WIDER conference on the New Economy in Development, May 10–11, 2002, Helsinki.

Clarke, G. 2001. "Bridging the Digital Divide: How Enterprise Ownership and Foreign Competition Affect Internet Access in Eastern Europe and Central Asia." World Bank Policy Research Working Paper 2629, Washington, DC.

Clarke, G., and S. Wallsten. 2004. "Has the Internet Increased Trade? Evidence from Industrial and Developing Countries." World Bank Policy Research Working Paper 3215, Washington, DC.

Cornford, J. 2001. "The Evolution of the Information Society and Regional Development in Europe." Mimeo, University of Newcastle, Newcastle, UK.

Crafts, N. 1999. "Development History." Paper presented at "The Future of Development Economics in Perspective" conference, Dubrovnik, May 1999.

———. 1998. "East Asian Growth Before and After the Crisis." IMF Working Paper No. 98/137, Washington, DC.

Croft, Layton. 2001. "Closing the Gap: Multi-Media Rice Information Service Gives Competitive Edge to Mongolian Nomads." Submission to infoDev/IICD Stories Project Website, http://www.iicd.com.

CSIR. 1998. "Knowledge in Development: Multi-Media, Multi-Purpose Community Information Centers as Catalysts for Building Innovative Knowledge-Based Societies." Background paper for the 1998 *World Development Report*, the World Bank, Washington, DC.

Daisey, M. 2002. *21 Dog Years: Doing Time @ Amazon.com*. New York: Free Press.

Daly, J. 2000. "Will the Internet Promote Democracy?" *IMP*, September, http://www.cisp.org/imp/september_2000/daly/09_00daly.htm.

Daveri, F., and O. Silva. 2004. "Not Only Nokia: What Finland Tells Us About New Economy Growth." *Economic Policy*, April, 117–163.

David, P. 2000. "Digital Technology and the Productivity Paradox: After Ten Years, What Has Been Learned." Mimeo, Stanford University, Stanford, CA.

David, P. 1990. "The Dynamo and the Computer: An Historical Perspective on the Modern Productivity Paradox." *The American Economic Review* 80, no. 2: 355–361.

Davidson, B., and B. Munslow. 1990. "The Crisis of the Nation-State in Africa." *Review of African Political Economy*, Winter: 9–21.

Davis, R. A. 2001. "Internet Abuse in the Workplace." http://www.internetaddiction.ca/cyberslacking.htm.

D'Costa, A. 2003. "Uneven and Combined Development: Understanding India's Software Exports." *World Development* 31, no. 1: 211–226.

Dedrick, J., and K. Kraemer. 2000. "Asia E-Commerce Report 2000." Mimeo, Center for Research on Information Technology and Organizations, University of California, Irvine.

———. 1998. *Asia's Computer Challenge: Threat or Opportunity for the United States and the World?* New York: Oxford University Press.

De Melo, J. 2000. *Telecommunications and the Poor*. Mimeo, World Bank, Washington, DC.

de Moura Castro, C., L. Wolff, and N. Garcia. 1999. "Bringing Education by Television to Rural Areas." *TechKnowLogia*, September–October. Available at http://www.techknowlogia.org.

Dertouzos, M. 1999. "The Rich People's Computer?" *MIT Technology Review*, January.

———. 1998. *What Will Be: How the New World of Information Will Change Our Lives.* San Francisco: HarperEdge.

Diamond, Jared. 1997. *Guns, Germs, and Steel: The Fates of Human Societies.* New York: W. W. Norton.

Doms, M., T. Dunne, and K. Troske. 1997. "Workers Wages and Technology." *Quarterly Journal of Economics* 112, no. 1: 253–290.

Dot-Com Alliance. 2005. "Powering ICT Toolkit." Available at www.dot-com-alliance.org, accessed June 23, 2005.

Duncombe, Richard, and Richard Heeks. 1999. "Information, ICTs and Small Enterprises: Findings from Botswana." Development Informatics Working Paper Series, Paper No. 7, IDPM, June, Manchester, UK.

Dyson, Freeman. 1999. *The Sun, the Genome, and the Internet.* New York: Oxford University Press.

Edwards, S. 1993. "Trade Policy, Exchange Rates and Growth," NBER Working Papers 4511, Cambridge, MA.

Egleston, K., R. Jensen, and R. Zeckhauser. 2001. "Information and Communications Technologies: Markets and Economic Development," in G. Kirkman, ed., *Global Information Technology Report.* New York: Oxford University Press.

EIU (Economist Intelligence Unit). 2002. *Chile: Country Report.* London: EIU.

Eltzroth, C., and C. Kenny. 2003. *Broadcasting and Development: Options for the World Bank.* Washington, DC: World Bank.

EMarketer. 2003. "An Elephant in the Room: The Online at-Work Audience." www.emarketer.com.

Eurostat. 2002. *E-Commerce in Europe.* Brussels, Belgium: Eurostat.

Fallows, J. 2000. "Technology We Hate." *American Prospect* 11, no. 6.

Farivar, C. 2005. "Waiting for That $100 Laptop? Don't Hold Your Breath." *Slate,* November 29, www.slate.com/id/2131201/.

Feldstein, M. 2003. "Why Is Productivity Growing Faster?" NBER Working Paper 9530, Cambridge, MA.

Felipe, Jesus. 1999. "Total Factor Productivity in East Asia: A Critical Survey." *Journal of Development Studies* 35, no. 4: 831–858.

Fink, C., A. Mattoo, and R. Rathindran. 2002. "An Assessment of Telecommunications Reform in Developing Countries." World Bank Policy Research Working Paper 2909, Washington, DC.

Fink, C., and K. Maskus, eds. 2005. *Intellectual Property and Development: Lessons from Recent Economic Research.* Washington, DC: World Bank.

Fogel, R. W. 1964. *Railroads and Economic Growth: Essays in Economic History.* Baltimore: Johns Hopkins University Press.

Forestier, E., J. Grace, and C. Kenny. 2001. "Can Information and Telecommunications Technologies Be Pro-Poor?" *Telecommunications Policy* 26, no. 11: 623–646.

Franda, M. 2002. *Launching into Cyberspace: Internet Development and Politics in Five World Regions.* Boulder, CO: Lynne Rienner.

G8. 2000. *Okinawa Charter on Global Information Society.* Tokyo: Government of Japan.

Galbraith, J. 1989. *A History of Economics: The Past as the Present.* London: Penguin.

Gallup Organization. 1996. *The Gallup India Survey Report.* Available at www.gallup.com/poll/specialReport.

Garcia-Mila, T., and T. J. McGuire. 1992. "The Contribution of Publicly Provided Inputs to States' Economics." *Regional Science and Urban Economics* 22: 229–241.

Gates, W. H. 2000. Speech given at the Digital Dividends conference in Seattle, October 18. Available at http://www.microsoft.com/billgates/speeches/2000/10-18digital dividends.asp.

Geroski, P. 1999. "Models of Technology Diffusion." Center for Economic Policy Research Discussion Paper No. 2146. London, UK.

Ghosh, S., and A. Kraay. 2000. "Measuring Growth in Total Factor Productivity." *World Bank PREM Notes*, no. 42, September.

Gillespie, A., R. Richardson, and J. Cornford. 2001. "Regional Development and the New Economy." *EIB Papers* 6, no. 1: 109–131.

Goldman Sachs. 2000. *B2B—Just How Big Is the Opportunity.* Available at http://www.gs.com/hightech/research.

Gomez, R. 2000. "The Hall of Mirrors: The Internet in Latin America." *Current History* 99: 634.

Goolsbee, A. 2000. "In a World Without Borders: The Impact of Taxes on Internet Commerce." *Quarterly Journal of Economics,* May: 561–576.

Goolsbee, A., and J. Guryan. 2002. "The Impact of Internet Subsidies in Public Schools." NBER working paper series, No. 9090, National Bureau of Economic Research, Cambridge, MA.

Gordon, R. 2003. "Five Puzzles in the Behaviour of Productivity, Investment and Innovation: Draft Chapter for World Economic Forum." *Global Competitiveness Report,* 2003–2004.

———. 2002. "Technology and Economic Performance in the American Economy." NBER Working Paper No. 8771, Cambridge, MA.

———. 2000a. "Does the New Economy Measure Up to the Great Inventions of the Past?" *Journal of Economic Perspectives* 14, no. 4: 49–74.

———. 2000b. "Interpreting the 'One Big Wave' in US Long-Term Productivity Growth," in B. van Ark, S. Kuipers, and G. Kuper, eds. *Productivity, Technology and Economic Growth.* Amsterdam: Kluwer.

Government of Chile. 2002. "Acceso y Uso de las TIC en las Empresas Chilenas, Division Tecnologías de Información y Comunicación." Mimeo, Subsecretaria de Economia, Government of Chile.

Grace, J., and C. Kenny. 2003. "A Short Review of Information and Communications Technologies and Basic Education in LDCs." *International Journal of Educational Development* 23: 627–636.

Grace, J., C. Kenny, and C. Qiang. 2001. "Information and Communication Technologies and Broad-Based Development: A Partial Review of the Evidence." World Bank Working Paper no. 12, Washington, DC.

Graham, S. 2001. "Information Technologies and Reconfigurations of Urban Space." *International Journal of Urban and Regional Research* 25, no. 2: 405–410.

Greenan, N., J. Mairess, and A. Topiol-Bensaid. 2001. "Information Technology and Research and Development Impacts on Productivity and Skills," in M. Pohjola, ed., *Information Technology, Productivity and Economic Growth.* Oxford: Oxford University Press.

Guillen, M., and S. Suarez. 2001. "Developing the Internet: Entrepreneurship and Public Policy in Ireland, Singapore, Argentina and Spain." *Telecommunications Policy* 25: 349–371.

Halewood, N., and C. Kenny. 2005. "Young People and Communications Technologies." Background Paper for the 2007 *World Development Report.* Mimeo, World Bank, Washington, DC.

Hamelink, C. 1999. "Human Development," in UNESCO, ed., *UNESCO World Communications and Information Report, 1999–2000.* New York: UNESCO.

Heeks, R. 2002a. "Failures, Successes and Improvisation of Information Systems Projects in Developing Countries." Development Informatics Working Paper Series 11, IDPM, Manchester, UK.

———. 2002b. "eGovernment in Africa: Promise and Practice." iGovernment Working Paper 13, IDPM, Manchester, UK.

———. 1999a. "Information and Communication Technologies: Poverty and Development." Development Informatics Working Paper Series No. 5, IDPM, Manchester, UK.

———. 1999b. *Information Technology, Government and Development.* Report on the IT, Government and Development workshop, Manchester, England, November 26. Mimeo, University of Manchester.

———. 1998. "Information Age Reform of the Public Sector: The Potential and Problems of IT for India." Information Systems for Public Sector Management Working Paper Series No. 6, IDPM, Manchester, UK.

Heeks, R., and C. Kenny. 2001. "Is the Internet a Technology of Convergence or Divergence?" Mimeo, World Bank, Washington, DC.

Heeks, R., and B. Nicholson. 2002. "Software Export Success Factors and Strategies in Developing and Transitional Economies." University of Manchester Development Informatics Working Paper Series No. 12, Manchester, England.

Hekl, J., and C. Waack. 2001. "Via Sebrae: An E-Commerce Solution for Small Business in Brazil." World Resources Institute Digital Dividend Case Study, World Resources Institute, Washington, DC.

Hilbert, M. 2001. *Latin America on Its Path into the Digital Age: Where Are We?* Santiago, Chile: CEPAL/ECLAC.

Hirsch, A. 1998. "Computer Training and Internet Access Issues in Wa, Upper West Province, Ghana." Mimeo, World Bank, Washington, DC.

Hirschman, A. 1967. *Development Projects Observed.* Washington, DC: Brookings Institution.

Hobday, M. 2001. "The Electronics Industries of the Asia Pacific." *Asia-Pacific Economic Literature* 15, no. 1: 13–29.

Howard, M. 1993. *The Lessons of History.* Oxford: Oxford University Press.

Hummels, D. 2001. "Time as a Trade Barrier." Mimeo, University of Chicago.

———. 1999. "Have International Transportation Costs Declined?" Mimeo, University of Chicago.

Humphrey, J., R. Mansell, D. Pare, and H. Schmitz. 2003. *The Reality of E-Commerce with Developing Countries.* Brighton: Institute of Development Studies.

ILO. 2001. *World Employment Report.* Geneva: ILO.

Indjikian, R., and D. Siegel. 2005. "The Impact of Investment in IT on Economic Performance: Implications for Developing Countries." *World Development* 33, no. 5: 681–700.

ITU. 2002. *International Telecommunications Indicators.* Geneva: ITU.

———. 2000. *International Telecommunications Indicators.* Geneva: ITU.

———. 1999. *Challenges to the Network: Internet for Development.* Geneva: ITU.

———. 1998a. "Rural Telecommunications in Colombia—Lessons Learned." Report prepared for the World Telecommunication Development Conference, WTDC-98, Valletta, Malta, March 23–April 1.

———. 1998b. *WTDC Backgrounder.* World Telecommunication Development Conference, WTDC-98, Valletta, Malta, March 23–April 1.

John Bartholomew and Son. 1989. *Atlas of the World.* London: John Bartholomew and Son.

Joseph, K. 2002. "Growth of ICT and ICT for Development: Realities and Myths from the Indian Experience." Paper presented at the WIDER Conference on The New Economy in Development, May 10–11, 2002, Helsinki.

Jupiter Research. 2005. *Municipal Wireless: Partner to Spread Risks and Costs While Maximizing Benefit Opportunities.* New York: Jupiter Research.

Keely, L., and D. Quah. 1998. "Technology and Economic Growth." Mimeo, LSE, London.

Kenny, C. 2005. "Why Are We So Worried About Income? Nearly Everything that Matters Is Converging." *World Development* 33, no. 1: 1–19.

———. 2003. "The Internet and Economic Growth in Developing Countries: A Case of Managing Expectations?" *Oxford Development Studies* 31, no. 1: 99–113.

———. 2002a. "Information and Communications Technologies for Direct Poverty Relief: Costs and Benefits." *Development Policy Review,* May, 20, no. 2: 141–157.

———. 2002b. "Should We Try to Bridge the Digital Divide?" *info* 4, no. 3: 4–10.

———. 2001. "Prioritizing Countries for Assistance to Overcome the Digital Divide." *Communications and Strategies,* no. 41: 17–34.

———. 1999. "Why Aren't Countries Rich? Weak States and Bad Neighborhoods." *Journal of Development Studies* 35, 5: 26–47.

Kenny, C., and D. Williams. 2001. "What Do Economists Know About Economic Growth—or Why Don't They Know Very Much?" *World Development* 29, no. 1: 1–22.

Kenny, C., B. Lanvin, and A. Lewin. 2003. "The Access Divide," in World Bank, ed., *ICT and Development: Enabling the Information Society.* Washington, DC: World Bank.

Kenny, C., I. Neto, S. Janakiram, and C. Watt. 2005. "The Bumpy Road to E-development?" in R. Schware, ed., *E-Development: From Excitement to Efficiency.* Washington, DC: World Bank.

Keremane, R., and C. Kenny. 2005. "Toward Universal Telephone Access: Market Progress and Progress Beyond the Market." Mimeo, World Bank, Washington, DC.

Khoung, V. 2004. "ICT and Economic Growth Across Countries: Contribution, Impact and Policy Implications." Ph.D. diss., Harvard University.

Kirkman, G., P. Cornelius, J. Sachs, and K. Schwab, eds. 2002. *The Global Information Technology Report, 2001–2002.* New York: Oxford University Press.

Klenow, P., and Rodriguez Clare, A. 1997. "Economic Growth: A Review Essay." *Journal of Monetary Economics* 40: 597–617.

Koch, J., and R. Cebula. 2002. "Price, Quality and Service on the Internet: Sense and Nonsense." *Contemporary Economic Policy* 20, no. 1: 25–37.

Kolko, J. 2002. "Silicon Mountains, Silicon Molehills: Geographic Concentration and Convergence of Internet Industries in the US." *Information Economics and Policy* 14: 211–232.

Kozma, R., et al. 1999. *World Links for Development: Accomplishments and Challenges.* Monitoring and Evaluation Report 1998–1999. Washington, DC: World Bank.

Kraut, R., S. Kiesler, B. Boneva, J. Cummings, V. Helgeson, and A. Crawford. 2002. "Internet Paradox Revisited." *Journal of Social Issues* 58, no. 1: 49–74.

Kraut, R., V. Lundmark, M. Patterson, S. Kiesler, T. Mukopadhyay, and W. Scherlis. 2000. "Internet Paradox: A Social Technology That Reduces Social Involvement and Psychological Well-Being?" *American Psychologist* 53, no. 9: 1017–1031.

Kumar, N., and K. Joseph. 2005. "Export of Software and Business Process Outsourcing from Developing Countries: Lessons from the Indian Experience." *Asia-Pacific Trade and Investment Review* 1, no. 1: 91–110.

Kumar, R. 2004. "eChoupals: A Study on the Financial Sustainability of Village Internet Centers in Rural Madhya Pradesh." *Information Technologies and International Development* 2, no. 1: 47–73.

Kunde, B. 1996. "A Brief History of Word Processing." Mimeo, Stanford University, Stanford, CA.

Kurlantzick, J. 2004. "The Web Won't Topple Tyranny." *The New Republic,* April 5, 2004.

Lal, B. 1999. "Information and Communication Technologies for Improved Governance." Mimeo, IICD, Amsterdam.

Lal, K. 1996. "Information Technology, International Orientation and Performance: A Case Study of Electrical and Electronic Goods Manufacturing Firms in India." *Information Economics and Policy* 8: 269–280.

Law Journal Extra. 1996. "A History of GATT and the Structure of the WTO." *International Contract Adviser* 2, no. 1.

Lee, I., and Y. Khatri. 2003. "Information Technology and Productivity Growth in Asia." IMF Working Paper WP/03/15, Washington, DC.

Lehr, B., and F. Lichtenberg. 1999. "Information Technology and Its Impact on Productivity: Firm-Level Evidence from Government and Private Data Sources, 1977–1993." *Canadian Journal of Economics* 32, no. 2: 335–362.

Lewis, M. 1999. *The New, New Thing.* London: Hodder and Stoughton.

Linton, F., A. Charron, and D. Joy. 1998. "OWL: A Recommender System for Organization-Wide Learning." MITRE Technical Report, http://www.mitre.org/technology/tech_tats/modeling/owl/owl.html.

Lobo, A., and S. Balakrishnan. 2002. "Report Card on Service of Bhoomi Kiosks: An Assessment of Benefits by Users of the Computerized Land Records System in Karnataka." Mimeo, Public Affairs Center, Bangalore.

Madani, D. 2000. "A Review of the Role and Impact of Export Processing Zones." Mimeo, World Bank, Washington, DC.

Maddison, A. 1995. *Monitoring the World Economy, 1820–1992.* Paris: OECD.

Madrick, J. 2001. "Lest We Forget." *New York Times*, May 10, 2001.

Malloch Brown, Mark. 2000. "Editorial." *Choices,* June (UNDP).

Malone, M. 1998. "Forget Digital Utopia. . . . We Could Be Headed for Technofascism." *Upside,* August.

Mann, C., S. Eckert, and S. Knight. 2000. *Global Electronic Commerce.* Washington, DC: International Institute for Economics.

Matthews, J. A., and D. Cho. 2000. *Tiger Technology: The Creation of a Semiconductor Industry in East Asia.* Cambridge: Cambridge University Press.

Mattoo, A. and L. Schuknecht. 2000. "Trade Policies for Electronic Commerce." Mimeo, World Bank, Washington, DC.

Maugis, V., N. Choucri, S. Madnick, M. Siegel, S. Gillett, F. Haghseta, H. Zhu, H. Best, and M. Best. 2003. "Global E-Readiness—for What? Readiness for E-Banking," MIT-Sloan Working Paper 4487-04.

Maurer, R. 1994. "International Trade and Economic Growth: A Survey of Empirical Studies." Kiel Working Paper No. 660. Kiel, Germany.

Mayer, J. 2000. "Globalization, Technology Transfer and Skill Accumulation in Low-Income Countries." WIDER Working Paper No. 150. Helsinki, Finland.

McKinsey and Company. 2001. *US Productivity Growth, 1995–2000.* Boston: McKinsey.

Melchor, A. 2001. "Global Income Inequality: Beliefs, Facts and Unresolved Issues." *World Economics* 2, no. 3: 87–108.

Miller, R. 2001. "Leapfrogging? India's Information Technology Industry and the Internet." IFC Discussion Paper No. 42. Washington, DC.

Mody, A. 1997. *Infrastructure Strategies in East Asia: The Untold Story.* Washington, DC: World Bank.

Molla, A. 2005. "Exploring the Reality of E-Commerce Benefits Among Businesses in a Developing Country." IDPM Manchester Development Informatics Working Paper Series No. 22, Manchester, England.

Moore, T. 1996. "Utopia," in S. Bruce, ed., *Three Early Modern Utopias.* Oxford: Oxford University Press.

Morrison, C., and D. Siegel. 1997. "External Capital Factors and Increasing Returns in US Manufacturing," *The Review of Economics and Statistics* 79, no. 4: 647–654.

Munnell, Alicia H. 1992. "Policy Watch: Infrastructure Investment and Economic Growth." *Journal of Economic Perspectives* 6: 189–198.

Myers, D. 2002. "Does Economic Growth Improve Human Morale?" *Quarterly Newsletter of the Center for a New American Dream,* March 20, 2002.

Naughton, K. 1999. "CyberSlacking." *Newsweek,* November 29, pp. 62–65.

Negroponte, N. 1998a. "One-Room Schools." *Wired,* September 6.

———. 1998b. "The Third Shall Be First: The Net Leverages Latecomers in the Developing World." *Wired,* January.

———. 1995. "Being Digital: A Book Preview." *Wired*, February 1.

Nelson, R. 1998. "Technological Advance and Economic Growth." Paper presented at the Forum on Harnessing Science and Technology for America's Economic Future, National Academy of Sciences, Washington, DC, February 2–3.

Newman, N. 1997. "From MSWord to MSWorld: How Microsoft Is Building a Global Monopoly." A NetAction White Paper, http://www.netaction.org.

Nezu, R. 2000. "E-Commerce: A Revolution with Power." Mimeo, OECD, Paris.

Nie, N., and L. Erbring. 2001. "Internet and Society: A Preliminary Report." Mimeo, Stanford Institute for the Quantitative Study of Society, UCLA.

Noble, D. 1999. *The Religion of Technology: The Divinity of Man and the Spirit of Invention.* New York: Penguin.

North, D. 1992. "Transactions Costs, Institutions and Economic Performance." Mimeo, International Center for Economic Growth, San Francisco, California.

OECD. 2004a. "*The Economic Impact of ICT: Measurement, Evidence and Implications* Paris: OECD.

———. 2004b. "*Information Technology Outlook* Paris: OECD.

———. 2000a. *Literacy in the Information Age: Final Report of the International Adult Literacy Survey.* Paris: OECD.

———. 2000b. *A New Economy? The Changing Role of Innovation and Information Technology in Growth.* Paris: OECD.

OECD Working Party on Telecommunications and Information Services Policies. 2004. "The Development of Broadband Access in Rural and Remote Areas." OECD Report DSTI/ICCP/TISP(2003)7/Final, Paris.

Oppenheimer, T. 1997. "The Computer Delusion." *Atlantic Monthly,* July, 45–62.

Ostrov, B. 2002. "Hong Kong's CyberPort: Do Government and High Tech Mix?" *Independent Review* 7, no. 2: 221–236.

Oxfam. 2000. *Education Now: Breaking the Cycle of Poverty.* Oxford: Oxfam.

Oxley, J., and B. Yeung. 2000. "E-Commerce Readiness: Institutional Environment and International Competitiveness." Mimeo, University of Michigan Business School. Ann Arbor, MI.

Patrinos, H. A. 2001. "Education Premiums and the Impact of Technology." Mimeo, World Bank, Washington, DC.

Perdue, L. 2001. "Will the Religious Right Make the Tech Slump Even Worse?" http://www.eroticabiz.com.

Perkins, A., and M. Perkins. 1999a. *The Internet Bubble: Inside the Overvalued World of High-Tech Stocks—and What You Need to Know to Avoid the Coming Shake-out.* New York: HarperBusiness.

———. 1999b. "Eyes Wide Shut." *Financial Planning,* December 1.

Pigato, M. 2001. "Information and Communication Technology, Poverty and Development in Sub-Saharan Africa and South Asia." Mimeo, World Bank, Washington, DC.

Pohjola, M. 2001. "Introduction," in M. Pohjola, ed., *Information Technology, Productivity and Economic Growth.* Oxford: Oxford University Press.

———. 1998. "Information Technology and Economic Development: An Introduction to the Research Issues." UNU WIDER Working Paper 153, Helsinki, Finland.

Potashnik, M., and D. Adkins. 1996. "Cost Analysis of Information Technology Projects in Education: Experiences from Developing Countries." *World Bank Education and Technology Series* 1, no. 3.

Potashnik, Michael, and Joanne Capper. 1998. "Distance Education: Growth and Diversity." *Finance and Development,* March, 42–45.

Pritchett, L. 2002. "The Tyranny of Concepts: CUDIE (Cumulated, Depreciated, Investment Effort) Is Not Capital." *Journal of Economic Growth* 5, no. 4: 361–384.

———. 1997. "Divergence, Big Time." *Journal of Economic Perspectives* 11: 3–17.

———. 1996a. "Mind Your p's and q's: The Cost of Public Investment Is Not the Value of Public Capital." World Bank Policy Research Working Paper 1660,Washington, DC.

———. 1996b. "Where Has All the Education Gone?" World Bank Policy Research Working Paper No. 1581, Washington, DC.

Putnam, R., with R. Leonardi and R. Nanetti. 1993. *Making Democracy Work: Civic Traditions in Modern Italy.* Princeton: Princeton University Press.

Pyramid Research. 2001. "Broadband: Leaps in Asia." *Pyramid Research Perspective* October 17, 2001.

Qiang, C., G. Clarke, and N. Halewood. 2005. "The Role of ICT in Doing Business," in C. Qiang, ed., *Information and Communications for Development: Trends and Policies.* Washington, DC: World Bank.

Qiang, C, A. Pitt, and S. Ayers. 2003. "The Contribution of Information Communication Technologies to Growth." World Bank Working Paper 24, Washington, DC.

Quah, D. 2002. "Technology Dissemination and Economic Growth: Some Lessons for the New Economy." Mimeo, LSE, London.

Richardson, R., and A. Gillespie. 2003. "The Call of the Wild: Call Centres and Economic Development in Rural Areas." *Growth and Change* 34, no. 1: 87–108.

Rifkin, J. 2001. *The Age of Access: The New Culture of Hypercapitalism, Where All of Life Is a Paid-For Experience.* New York: Tarcher/Putnam.

Roberti, M. 2001. "General Electric's Spin Machine." *Industry Standard,* January 22–29.

Robinson, S. 2000. *Telecentres in Mexico: Learning the Hard Way,* presented at Partnerships and Participation in Telecommunications for Rural Development: Exploring What Works and Why, conference at the University of Guelph, Guelph, Ontario, Canada, October 26–27.

Rodriguez, F., and D. Rodrik. 1999. "Trade Policy and Economic Growth: A Skeptics Guide to the Cross-National Evidence." Mimeo, University of Maryland, College Park.

Rodriguez, F., and E. Wilson. 2000. "Are Poor Countries Losing the Information Revolution?" *Info*Dev Working Paper, Washington, DC.

Sachs, J., and A. Warner. 1997. "Sources of Slow Growth in African Economies." *Journal of African Economies* 6, 333–376.

Schiller, R. 2000. *Irrational Exuberance*. Princeton, NJ: Princeton University Press.

Schreyer, P. 2000. "The Contribution of Information and Communications Technology to Output Growth: A Study of the G7 Countries." OECD, STI Working Paper DSTI/DOC(2000)2, Paris.

Schware, R. 2004. "Options for Structuring a Social-e-Applications Rollout Venture Fund." Mimeo, World Bank, Washington, DC.

Schware, R., and A. Dean. 2003. "Deploying e-Government Programs: The Strategic Importance of 'I' Before 'E.'" *info* 5, 4: 10–19.

Schware, R., and S. Hume. 1996. "Prospects for Information Services Exports from the English-Speaking Caribbean." Mimeo, World Bank, Washington, DC.

Seligman, B. 1966. *Most Notorious Victory: Man in an Age of Automation*. New York: Free Press.

Shakeel, H., M. Best, B. Miller, and S. Weber. 2001. "Comparing Urban and Rural Telecenters' Cost." *Electronic Journal on Information Systems in Developing Countries* 4, no. 2: 1–13.

Slater, D., and J. Kwami. 2005. "Embeddedness and Escape: Internet and Mobile Use as Poverty Reduction Strategies in Ghana." ISRG Working Paper, London School of Economics, June.

Smith, A. 1976. *An Enquiry into the Nature and Causes of the Wealth of Nations*. Oxford: Oxford University Press.

Souter, D. 2005. *The Economic Impact of Telecommunications on Rural Livelihoods and Poverty Reduction*. London: Commonwealth Telecommunications Organisation.

Srinivasan, T. 2005. "Information Technology Enabled Services and India's Growth Prospects." Mimeo, Yale University, New Haven, CT.

Stiroh, K. 2002a. "Information Technology and the US Productivity Revival: What Do the Industry Data Say?" *American Economic Review*, December.

———. 2002b. "Are ICT Spillovers Driving the New Economy?" *Review of Income and Wealth* 48, no. 1: 33–57.

———. 1998. "Computers, Productivity and Input Substitution." *Economic Enquiry* 36, no. 2: 175–191.

Strassman, P. 2000. "The Perverse Economics of Information." *Impacts Magazine*, September.

Subramanian, A. 2004. "Medicines, Patents and TRIPS." *Finance and Development*, March.

Sudan, R. 2005. "The Basic Building Blocks of E-Government," in R. Schware, ed., *E-Development: From Excitement to Efficiency*. Washington, DC: World Bank.

Triplett, J. 1999. "The Solow Productivity Paradox: What Do Computers Do to Productivity?" *Canadian Journal of Economics* 32, no. 2: 309–334.

UCLA Center for Communication Policy. 2001. *UCLA Internet Report: Surveying the Digital Future*. Mimeo, UCLA Center for Communication Policy, Los Angeles.

UK Department for Trade and Industry. 2000. *Moving into the Information Age: An International Benchmarking Study*. London: Department for Trade and Industry.

UNCTAD. 2003a. *E-Commerce and Development Report, 2003*. New York: United Nations.

———. 2003b. *Information and Communication Technology Development Indices*. New York: United Nations.

———. 2001. *E-Commerce and Development Report*. Geneva: United Nations.

UNCTAD and ICTSD. 2001. "Intellectual Property Rights and Development." Policy Discussion Paper, UNCTAD, New York.

UNDP. 2001. *Human Development Report.* New York: UNDP.

———. 1999. *Human Development Report, 1999.* New York: Oxford University Press.

UNDP Evaluation Office. 2001. "Information Communications Technology for Development." *Essentials: Synthesis of Lessons Learned* 5, September, 12.

UNESCAP. 1999. *Economic and Social Survey of Asia and the Pacific, 1999.* Bangkok: UNESCAP.

US Census Bureau. 2005. *US Department of Commerce E-Stats, 2005.* www.census.gov/estats.

US Department of Commerce. 2000. *Falling Through the Net: Toward Digital Inclusion.* Washington, DC: Department of Commerce.

———. 1999. *The Emerging Digital Economy II.* Washington, DC: Department of Commerce.

van Ark, B., and M. Piatowski. 2004. "Productivity, Innovation, and ICT in Old and New Europe," *International Economics and Economic Policy* 1, nos. 2–3: 215–246.

Venables, A. 2001. "Geography and International Inequalities: The Impact of New Technologies." Mimeo, London School of Economics.

Wade, R. 2001. "How to Harness Information and Communication Technologies (ICT) for Wealth Creation in Developing Countries, and What Donors Can Do to Help. Or, When Are Development Fads Beneficial?" Paper prepared for OECD/UNDP Conference on the Knowledge Economy, OECD, Paris, March 5–6, 2001.

Walde, K., and C. Wood. 1999. "The Empirics of Trade and Growth: Where Are the Policy Recommendations?" Mimeo, World Bank, Washington, DC.

Wallsten, S. 2005. "Broadband Penetration: An Empirical Analysis of State and Federal Policies." AEI-Brookings Joint Center Working Paper 05-12, Washington, DC.

———. 2001. "International Telecommunications Accounting Rate Reforms: The Effects of Reduced Rates on Telecom Traffic and Investment in Developing Countries." *Journal of Regulatory Economics* 20, no. 3: 307–323.

———. 1999. "An Empirical Analysis of Competition, Privatization and Regulation in Africa and Latin America." Mimeo, World Bank, Washington, DC.

Warschauer, M. 2002. "Reconceptualizing the Digital Divide." *First Monday* 7, no. 7. www.firstmonday.org.

Weisburd, D., C. Lum, and A. Petrosino. 2001. "Does Research Design Affect Study Outcomes in Criminal Justice?" *ANNALS of the AAPSS* 578, November.

Wellenius, B. 2005. "Extending Access to Communication and Information Services: Guiding Principles and Practical Solutions," in Qiang, C., ed., *Global ICT Development Report.* World Bank: Washington, DC.

Wheeler, D. 2004. *Chile: New Economy Study.* Washington, DC: World Bank.

———. 2000. "A Research Contribution to the World Bank's ICT Sector Strategy." Mimeo, World Bank, Washington, DC.

White House, Office of the Press Secretary. 2000. *Information Technology Research and Development: Information Technology for the Twenty-First Century.* January 21.

Wong, W., and E. Welch. 2004. "Does E-Government Promote Accountability? A Comparative Analysis of Website Openness and Government Accountability." *Governance* 17, 2: 275–298.

World Bank. 2005. *Financing Information and Communication Infrastructure Needs in the Developing World: Public and Private Roles.* Washington, DC: World Bank.

———. 2002a. *Global Economic Prospects and the Developing Countries.* Washington, DC: World Bank.

———. 2002b. *World Development Indicators.* Washington, DC: World Bank.

———. 2001a. *World Development Indicators.* New York: Oxford University Press.

———. 2001b. *World Development Report.* New York: Oxford University Press.

———. 2001c. *Republic of Tunisia: Information and Communications Technology Strategy Report.* Washington, DC: World Bank.

———. 2000a. *Global Economic Prospects.* Washington, DC: World Bank.

———. 2000b. *The Networking Revolution: Opportunities and Challenges.* Washington, DC: World Bank.

———. 2000c. *World Development Report, 1999–2000: Entering the Twenty-First Century.* Oxford: Oxford University Press.

———. 1999. *World Development Report.* Washington, DC: World Bank.

———. 1996. *Tanzania: The Challenge of Reforms, Growth, Income and Welfare.* World Bank, Washington, DC.

Wright, David. 1997. *Telemedicine and Developing Countries: A Report of Study Group 2 of the ITU Development Sector.* Geneva: ITU.

Wyatt, S. 1999. "They Came, They Surfed, They Went Back to the Beach: Why Some People Stop Using the Internet." Paper presented at the Society for Social Studies of Science Conference, San Diego, October 1999.

Young, A. 1995. "Growth Without Scale Effects." *Journal of Political Economy* 106: 41–63.

Yusuf, S. 2004. *Innovative East Asia: The Future of Growth.* New York: Oxford University Press.

Index

About the Book

OPINION LEADERS IN GOVERNMENT and business routinely tout the Internet's power as a force for economic and social development, and programs designed to bridge the digital divide are springing up across the developing world. Many questions remain, however, about the effectiveness of such programs in fostering greater productivity and improving quality of life. *Overselling the Web?* offers a much needed antidote to the Internet hype touting the promise of new technologies.

Drawing on macroeconomic data as well as eye-opening anecdotes, Charles Kenny underscores the trade-offs and constraints inherent in the new communications technology. His work raises serious questions about the advisability of channeling scarce investment funds into the Internet when countries are confronting more basic challenges in the realm of education, health, and infrastructure.

Charles Kenny is a development economist based in Washington, DC. He has spent the past six years working on projects and research related to information and communications technologies in developing countries, including activities in Africa, Latin America, and Asia.

iPolitics: Global Challenges in the Information Age

Renée Marlin-Bennett, series editor

Governing the Internet:
The Emergence of an International Regime,
Marcus Franda

Launching into Cyberspace:
Internet Development and Politics in Five World Regions,
Marcus Franda

Knowledge Power: Intellectual Property, Information, and Privacy,
Renée Marlin-Bennett

Negotiating Privacy:
The European Union, the United States,
and Personal Data Protection,
Dorothee Heisenberg

Intellectual Property Rights: A Critical History,
Christopher May and Susan K. Sell

Negotiating the Net in Africa:
The Politics of Internet Diffusion,
edited by Ernest J. Wilson III and Kelvin R. Wong

Overselling the Web? Development and the Internet,
Charles Kenny